ANIMAL SCIENCE, ISSUES AND RESEARCH

CRUSTACEANS

PHYSIOLOGICAL CHARACTERISTICS, EVOLUTION AND CONSERVATION STRATEGIES

Animal Science, Issues and Research

Additional books in this series can be found on Nova's website under the Series tab.

Additional e-books in this series can be found on Nova's website under the eBooks tab.

ANIMAL SCIENCE, ISSUES AND RESEARCH

CRUSTACEANS

PHYSIOLOGICAL CHARACTERISTICS, EVOLUTION AND CONSERVATION STRATEGIES

VICTOR ALVARADO
EDITOR

NOVINKA

New York

Library of Congress Cataloging-in-Publication Data

ISBN: 978-1-63484-890-9

Published by Nova Science Publishers, Inc. † New York

CONTENTS

PREFACE

Crustaceans are key organisms in aquatic ecosystems. Today crustaceans of Ponto-Caspian origin have become important faunal constituents in many European water bodies, including the Baltic Sea. In this book, Chapter One reviews the causes, consequences and prospects of Pont-Caspian crustaceans in the Baltic Sea. Chapter Two examines the physiological causes and ecological implications of extra and "intermediate" larval stages in decapod crustacean life. Chapter Three discusses a comparative study of liposoluble vitamin effects in decapod crustaceans. The final chapter is a commentary on the possible management and conservation strategies of crustaceans from temperate sandy beaches.

In: Crustaceans
Editor: Victor Alvarado

ISBN: 978-1-63484-890-9

Chapter 1

PONTO-CASPIAN CRUSTACEANS IN THE BALTIC SEA: CAUSES, CONSEQUENCES, PROSPECTS

Aldona Dobrzycka-Krahel
Department of Experimental Ecology of Marine Organisms,
Institute of Oceanography, University of Gdansk, Poland

ABSTRACT

Crustaceans are key organisms in aquatic ecosystems. Today crustaceans of Ponto-Caspian origin have become important faunal constituents in many European water bodies, including the Baltic Sea. These species probably migrated actively and passively along water routes connecting the catchment areas of the Baltic Sea, as well as the Azov, Black and Caspian Seas. The spread of Ponto-Caspian species via water routes was likely facilitated by global warming and anthropogenic increases in the salinity of inland waters. The main sources of ion discharges (mainly chloride ions) were coal and salt mining, irrigation, vegetation clearance, dryland farming and industry in general. In this way Ponto-Caspian species were able to overcome geographical and "freshwater" barriers.

The living environment of the Baltic Sea is difficult because, for example, of the low salinity, changeable temperature, and the presence of hypoxia/anoxia conditions and hydrogen sulphide. Only very flexible, eurytopic species with a broad tolerance to a range of environmental factors can establish themselves there. One of the most important factors is salinity. This affects life in the specific environment of this sea containing the "Artenminimum" (native species minimum) zone in the 5-8 PSU salinity range. There are similar "Artenminimum" zones in the Sea of Azov and Caspian Sea. The 5-8 PSU salinity range acts not only as an ecological "frontier" but also as a physiological and evolutionary barrier. The low native species richness in the Baltic Sea and empty ecological niches and habitats are favourable to flexible alien species e.g., from the Ponto-Caspian region. Clearly, Ponto-Caspian crustaceans that have successfully established themselves in the Baltic Sea exhibit plasticity in their tolerance to environmental stress. The ability of organisms to colonize novel environments does not depend solely on their tolerance of environmental conditions but also on their abilities to function in new habitats beyond their native range. Fitness-related plasticity undergoes adaptive changes in response to the environment.

Colonization by non-native Ponto-Caspian crustaceans may have important consequences for organisms in the Baltic Sea. The populations of non-native gammarids, which have taken just a few years to become established, have completely supplanted the native gammarid species in the near-shore zones of the Vistula Lagoon. The retreat of native species is a serious problem in many places colonized by alien gammarids, not only in Poland, but right across Europe. The possible reasons behind this phenomenon include pollution of aquatic ecosystems to the extent that they are no longer tolerated by indigenous gammarids, predation pressure from the newcomers, and the plasticity of aliens together with their greater reproductive capacity. The changing environment of the Baltic Sea may affect the prospects for Ponto-Caspian crustaceans in this sea as a result not only of the displacement of native species, but also the appearance of other new species from the Ponto-Caspian region.

Keywords: Ponto-Caspian crustaceans, Baltic Sea, critical salinity, fitness

INTRODUCTION

Most Ponto-Caspian crustaceans are known to be recent invaders of European inland waters and are often referred to as "freshwater invaders" (Gherardi 2007, Gallardo & Aldridge 2013). In the recipient ecosystems, these species occupy mainly rivers, lakes and other inland water bodies (Mordukhai-Boltovskoi 1960, Dedyu 1980, Konopacka 1998, Bącela & Konopacka 2005, Casellato et al. 2006, 2008, Berezina 2007, Mastitsky & Makarevich 2007, Grabowski et al. 2007 b, Pöckl 2009, Tricarico et al. 2010). The transition from saline Ponto-Caspian waters to freshwater habitats was a dramatic shift between "adaptive zones" (Lee & Bell 1999). Freshwater invaders may originate from habitats with wide temporal or spatial ranges in salinity, where broad salinity tolerances can form. The preadaptation to functioning in waters of variable salinity was the wide fluctuation in the salinity of the Ponto-Caspian region itself; the geological, climatological and biogeographical history of this region influenced the adaptations of Ponto-Caspian fauna (Reid & Orlova 2002). The scale of salinity fluctuations in the native region could be an important factor in the acclimation response of species inhabiting this area (Lee & Bell 1999).

Many Ponto-Caspian species, particularly gammarids and mysids, have been deliberately transplanted into freshwater habitats in Eastern Europe and into artificial water bodies for fish enrichment purposes (Gasjunas 1972, Ioffe 1973). These artificial habitats may have facilitated their entry into inland waters. In any case, the anthropogenic increase in the salinity of inland waters enabled Ponto-Caspian crustaceans to spread towards European waters (Braukmann & Bohme 2011). As freshwater invaders, Ponto-Caspian crustaceans may acquire the ability to complete their entire life cycle in fresh water (Grabowski et al. 2007a). Today we are witnessing the transition from early established freshwater to saline habitats and hence the crossing of the next adaptive zone. Once these species had spread to more saline, brackish waters (including the Baltic Sea), scientists began to devote greater attention to the crossing of geographical and adaptive barriers by Ponto-Caspian crustaceans.

The Crossing of Barriers by Ponto-Caspian Crustaceans

The dispersal of Ponto-Caspian species may well have taken place by passive and active migration along waterways connecting the catchment areas of the Baltic Sea, as well as the Azov, Black and Caspian Seas (Bij de Vaate et al. 2002). Ships sailing these canals were probably the main transport vector. The spread of Ponto-Caspian species along waterways was probably facilitated by additional factors such as global warming and anthropogenic increases in the salinity of inland waters during the 20th century, because temperature and salinity are environmental factors that could raise a barrier to expansion. The main sources of ion discharges (mainly chloride ions) were coal and salt mining, irrigation, vegetation clearance, dryland farming and industry in general (e.g., Braukmann & Bohme 2011, Schulz 2011, Motyka & Postawa 2000, Korycińska & Królak 2006). In Europe, some inland water stretches have salinity levels similar to or even higher than those measured in the brackish Baltic Sea (Bäthe & Coring 2011, Petruk & Stoffler 2011). Moreover, freshwater habitats experience greater temperature fluctuations than do brackish or marine habitats. In such ways Ponto-Caspian species can overcome geographical and "freshwater" barriers.

The high tolerance of Ponto-Caspian fauna to varying environmental conditions and their ability to survive exposure to a range of salinities were the effect of the environmental conditions prevailing in the Ponto-Caspian region and in the Baltic Sea, and it is these factors that have probably contributed to the invasion success of these species (Reid & Orlova 2002). These opportunities acquired by Ponto-Caspian crustaceans over the course of evolution thus give them a head start when colonizing new areas. The highest numbers of individuals of these species in inland waters have been found in waters with elevated conductivities (e.g., Grabowski et al. 2009). Furthermore, the increasing salinity in the catchment area of the Vistula Lagoon resulting from industrial and agricultural activities there during the 20th century (Chubarenko & Tchepikova 2001) enabled Ponto-Caspian crustaceans to expand into this region. The successful establishment of alien gammarids in the zones of the Vistula Lagoon and the Vistula Delta was a stepping stone on their way to colonizing more saline waters, e.g., the brackish Baltic waters (Figure 1).

Salinity is also one of the first abiotic factors affecting species when they are transported in ballast waters. Most coastal ports are brackish habitats and a significant source of invasion (Paavola et al. 2005). The euryhalinity of many Ponto-Caspian crustaceans enables them to survive in ballast waters and colonize such brackish coastal port habitats.

Ponto-Caspian crustaceans crossed the critical salinity zones of the Baltic Sea as a result of their ability to function in such salinities.

Within the 5–8 PSU salinity range the number of species slopes to a minimum (Remane 1934) (Figure 2). Zenkevitch (1959) revealed similar Artenminimum zones in the Sea of Azov and Caspian Sea. This indicates that a critical salinity barrier does not exist for alien species already well adapted to the critical salinity range in their native areas.

Kinne (1971) has coined the term "horohalinicum" for this zone. The salinity range of 5–8 PSU acts as a "border" between saline and fresh water. Many properties of the water, such as ion concentrations and ratios, change in this zone. In sea waters ion ratios are roughly constant and the Knudsen rule enables the major ions and total salinity to be determined from a single ion concentration (Khlebovich 1968, 1974, 1989, 1990, Różańska 1987). In fresh waters, however, ion ratios vary considerably (Różańska 1987). The major anions in marine and fresh waters are Cl^- and HCO_3^- respectively, while the dominant cations are Ca^{2+} in typical fresh waters and Na+ in sea waters (Holland 1978). Apart from Cl^-, K^+ also has a major effect on aquatic organisms: it is the fourth most common cation after Na^+, Mg^{2+} and Ca^{2+} in the sea. The potassium concentrations in inland waters take a wide range of values. This element is the primary intracellular cation; it is also important in the activation of Na^+/K^+-ATPase (Mantel & Farmer 1983), a key component of extracellular volume regulation in the osmoregulation process. Hence, as enzyme activity can be directly related to the K^+ concentration (Bursey & Lane 1971), inadequate levels of aqueous K^+ are potentially detrimental with regard to the ability to effectively osmoregulate. Potassium ions are therefore highly significant for organism survival (Luke et al. 2007). Chloride (Cl^-) and potassium (K^+) concentrations were found to be the most closely linked to the mortality of *Dikerogammarus haemobaphes* from the Gulf of Gdańsk (the Polish part of the Baltic Sea, salinity 7 PSU) (Dobrzycka-Krahel & Graca 2014). It was found that not only the absolute ion concentrations are of major significance in the survival of aquatic organisms; the ratios of these ions are equally important (e.g., Fielder et al. 2001; Luke et al. 2007). The change in the ion ratio in water plays a decisive role in chemical processes within the salinity gradient. Experiments on *D. haemobaphes* (Dobrzycka-Krahel & Graca 2014) showed a strong connection between the Na^+/K^+ ratio and mortality in *D. haemobaphes*. In water of salinity 0.1 PSU, the Na^+/K^+ ratio was 10 (high mortality). There was zero mortality at salinity 7 PSU, where Na^+/K^+ was 42. This value is closest to the one prevailing in the Caspian Sea. Mortality of *D. haemobaphes* was also linked to the Na^+/Cl^- ratio: it was lowest (zero mortality) at $Na^+/Cl^- = 0.59$ (7 PSU), a value corresponding to that of the Caspian Sea. In water of salinity 0.1 PSU, this ratio was 64.67. In ocean waters this ratio is 0.56, but in the brackish Baltic and Caspian Seas, the values are higher at 0.60 and 0.59 respectively. Changes in the Cl^- concentration and Na^+/Cl^- in water are accompanied by changes in Na^+/Cl^- in the haemolymph of organisms (Sutcliffe 1971). As a result, the capacity of regulatory systems enabling organisms to adjust to different environments may be limited, with increased mortality as a consequence.

The salinity barrier of >6.5 PSU is not always overcome by organisms: for instance, the mussel *Dreissena polymorpha*, of Ponto-Caspian origin, does not reach Baltic Sea coastal waters. This species lives in brackish waters, for example, in the Szczecin Lagoon (Wolnomiejski & Woźniczka 2008), the Vistula Lagoon (Chubarenko & Margoński 2008), the Gulf of Finland (Orlova et al. 2004), the Curonian Lagoon (Chuševe et al. 2012) and Riga Bay (Kotta et al. 2008), but is unable to adapt to the salinity of the coastal southern Baltic zone, where the salinity is >6.5 PSU.

Figure 1. The Baltic Sea.

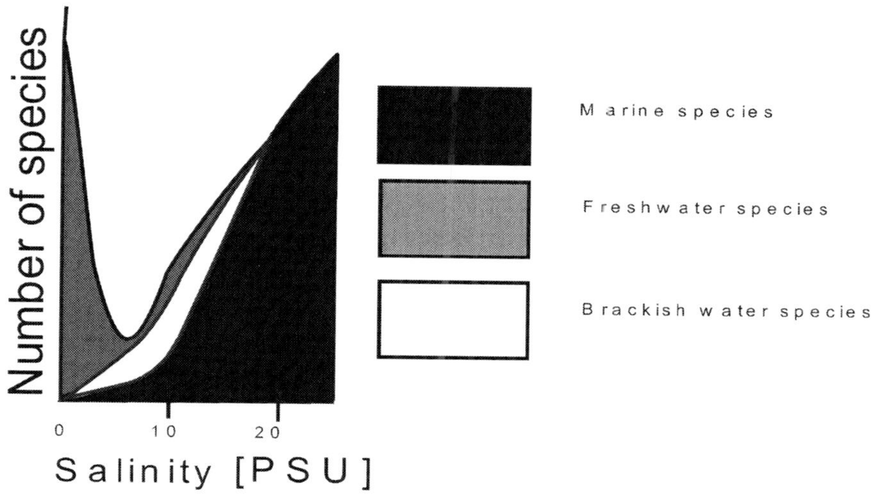

Figure 2. The modified Remane curve (after Remane, 1934).

PONTO-CASPIAN CRUSTACEANS IN THE BALTIC SEA ENVIRONMENT

The Ponto-Caspian gammarid species established in the Baltic Sea include *Pontogammarus robustoides* (G.O. Sars, 1894), *Obesogammarus cra*ssus (G.O. Sars, 1894), *Dikerogammarus haemobaphes* (Eichwald, 1841) (the demon shrimp), *Dikerogammarus villosus* (Sowinsky, 1894) (the killer shrimp), *Chaetogammarus (Echinogammarus) ischnus* (Stebbing, 1899), *Echinogammarus trichiatus* (Martynov, 1932) and *Chaetogammarus warpachowskyi* Sars, 1897.

Among the Corophiidae *Chelicorophium curvispinum* (G.O. Sars, 1895) (the Caspian mud shrimp) is of Ponto-Caspian species.

Ponto-Caspian mysid species recorded in the Baltic Sea include *Hemimysis anomala* G.O. Sars, 1907, *Paramysis lacustris* (Czerniavsky, 1882) and *Limnomysis benedeni* Czerniavsky, 1882.

Cercopagis pengoi (Ostroumov, 1891), the fishhook water flea, is a planktonic cladoceran of Ponto-Caspian origin. Other Ponto-Caspian planktonic crustaceans in the Baltic Sea are *Evadne anonyx* G.O. Sars, 1897 and *Cornigerius maeoticus* (Pengo, 1879).

The Ponto-Caspian isopod *Jaera istri* Veuille, 1979 is spreading towards the Baltic Sea along the River Oder.

One of the reasons for the fitness of Ponto-Caspian species is their very high productive potential, which is expressed in high fecundity, early maturity, length of breeding period and multiple reproduction cycles each year (Devin et al. 2004, Grabowski et al. 2007). As invaders of brackish waters, Ponto-Caspian crustaceans are able to complete their entire life cycle in brackish waters. Moreover, Ponto-Caspian crustaceans have often non-specific food preferences (Bij de Vaate et al. 2002), often being omnivorous.

Many species display a broad salinity tolerance, which facilitates their expansion (Kefford et al. 2004). The ability of organisms to colonize novel biotopes does not depend solely on their salinity tolerance but also on their ability to function at different salinities, that is, on their ability to actively regulate the osmotic concentration of their body fluids.

Although the high mortality of *D. haemobaphes* at an environmental salinity of 0.1 PSU, fresh or low-saline waters do not present an osmotic barrier either to *D. haemobaphes* or to other Ponto-Caspian gammarids like *P. robustoides, O. crassus* and *D. villosus* (Dobrzycka-Krahel & Surowiec 2011, Dobrzycka-Krahel & Graca 2014, Dobrzycka-Krahel et al. 2015, Dobrzycka-Krahel et al. under review, Dobrzycka-Krahel & Graca in preparation).

Osmoregulation is an excellent indicator of how external salinity can affect the homeostasis of an organism's internal medium. Many papers describe the strategies by which organisms use their osmoregulatory abilities in order to adjust to the environmental salinity (e.g., McLusky et al. 1982, Dobrzycka & Szaniawska 1995, Dobrzycka-Krahel & Surowiec 2011, Dobrzycka-Krahel et al. 2014, Hørlyck 1973). Osmoregulation determined in laboratory studies has shown that the Ponto-Caspian gammarids *P. robustoides, D. haemobaphes* and *D. villosus* can control their osmotic concentration over a wide range of salinity. The experimental salinities covered a range up to about 20 PSU since this is the highest value prevailing in Baltic Sea coastal waters.

Studied Ponto-Caspian gammarids are euryhaline species, in which osmoregulation is hyperosmotic within a salinity range from around 0 to 20 PSU (Dobrzycka-Krahel & Surowiec 2011, Dobrzycka-Krahel & Graca 2014, Dobrzycka-Krahel et al. 2015). The osmoregulatory curves demonstrate osmoregulatory abilities at such salinities.

A possible indicator of the adaptation of organisms is their fitness, as demonstrated in relationships between the length and weight of specimens. The total length - weight relationships for species calculated according to the power equation $y=ax^b$, ($W_w=aL_t^b$), where y (W_w) – wet weight in g, x (L_t) – total length in mm, a – intercept, b – slope. Species usually differ in their body weight to length proportions (Lindqvist & Lahti 1983), so their length-weight relationships vary. According to Hile (1936), however, the higher the value of slope 'b', the better the condition of individuals, whereas a low value of slope 'b' could mean a poorer individual condition and a smaller body weight in relation to length.

The length-weight relationships of Ponto-Caspian gammarids from the Gulf of Gdańsk and the Vistula Lagoon and Delta show that they are well adapted to such environmental conditions (Dobrzycka-Krahel et al. accepted, Dobrzycka-Krahel et al. under review). Good fitness corresponds to normal conditions and normal responses, and indicates that species have adapted well to the environment, which affords them excellent conditions for growth. The good adaptation of Ponto-Caspian crustaceans are also due to similarities between native and new regions, e.g., increasing urbanization, pollution and eutrophication due to industry, intensive agriculture, overexploitation of fish stocks and abundant sea traffic and port activities (Ojaveer et al. 2002).

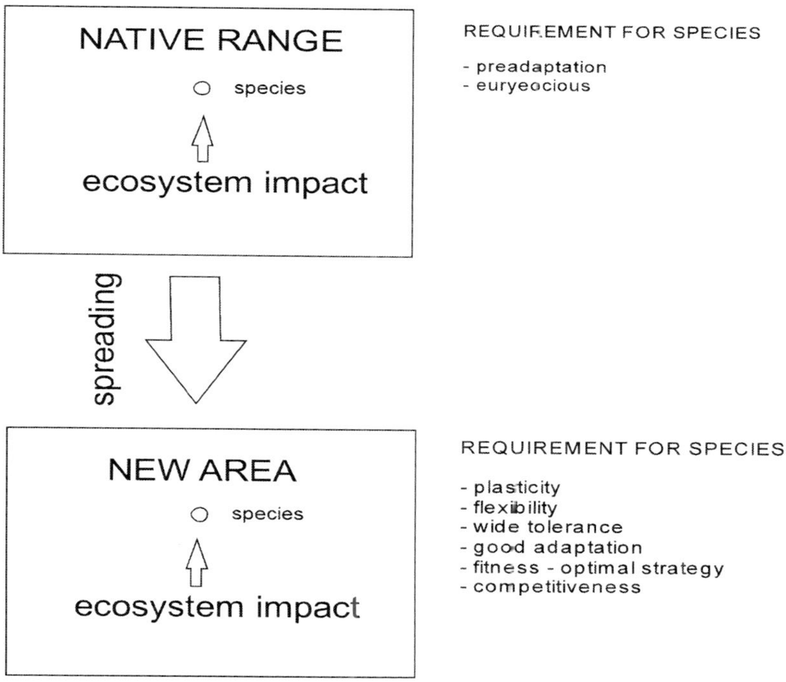

Figure 3. Establishment pattern of Ponto-Caspian crustaceans in the Baltic Sea.

On the other hand the fitness of alien species affects the process of invasion or colonization. The optimal strategy of aliens affects the colonization of the Baltic Sea. A good level of fitness can help Ponto-Caspian crustaceans maintain a strong competitive position in the environment. The fitness of a resident species increases its ability to oppose invaders, while the fitness of an invader enhances its ability to displace residents. Aliens incur only minor fitness costs by deviating from the optimum, allowing them to eventually gain a foothold. Deviations from this strategy will have opposing effects on overall fitness, which will determine the net effect (Peacor et al. 2006).

When an invasive species strays from its optimal strategy, it can no longer compete with native species. Since invaders did not undergo selection in the new environment, deviating from their foraging optimum could place them at a competitive fitness disadvantage (Gross 2006).

Consequences of the Presence of Non-Native Ponto-Caspian Crustaceans in the Baltic Sea

The colonization of non-native Ponto-Caspian crustaceans may have important consequences for the Baltic Sea fauna. Alien Ponto-Caspian invaders are strong competitors and have displaced native species. The populations of non-native gammarids, which have taken just a few years establish, have completely supplanted the native gammarid species in the Vistula Lagoon (Surowiec & Dobrzycka-Krahel 2008). The retreat of native species is a serious problem in many places colonized by alien amphipods, not only in Poland, but right across Europe. The possible reasons for this phenomenon include pollution of aquatic ecosystems to the extent that they are no longer tolerated by indigenous gammarids, predation pressure from the newcomers, the plasticity and the greater reproductive capacity of alien species. Some Ponto-Caspian crustaceans behave very aggressively towards native and other alien species. *Dikerogammarus villosus* is one of the most successful Ponto-Caspian immigrant thanks to its predatory habits towards invertebrates such as *Asellus aquaticus, Neomysis integer, Gammarus duebeni* and even *Gammarus tigrinus* (Dick & Platvoet 2000, Dick et al. 2002). It has been called the "killer shrimp," because it can eliminate other organisms. *Cercopagis pengoi*, the fishhook water flea, is a planktonic cladoceran crustacean that preys on native zooplankton, fish and fish roe. It causes economic losses (Jażdżewski & Grabowski 2011), fouls fishing nets and other equipment, and may have an array of other ecological impacts on aquatic systems. For these reasons, both species (*D. villosus* and *C. pengoi*) have been placed on the list of the 100 most invasive species in the DAISE (Delivering Alien Invasive Species Inventory for Europe) data base. Ponto-Caspian crustaceans are strong species and can play an important role in the sea. They have become dominant members of food webs in the Baltic Sea and significantly affected this ecosystem (Olenin & Leppäkoski 1999, Leppäkoski & Olenin 2000, Laxson et al. 2003, Ojaveer et al. 2004, Põllumäe & Kotta 2007).

Prospects for Ponto-Caspian Crustaceans

The changing environment of the Baltic Sea may affect the prospects for Ponto-Caspian crustaceans in this sea, with respect not only to the displacement of native species, but also to the appearance of other new species from the Ponto-Caspian region. The predicted desalinization of the Baltic Sea (Rajasilta et al. 2014, Vuarinen et al. 1998), caused by the lack of saline water pulses since the late 1970s and the simultaneous increase in run-off (Malmberg & Svanson 1982, Launiainen & Vilhma 1990; Matthäus & Schinke 1994), will have no effect on any future limitation on the occurrence of Ponto-Caspian crustaceans in the Baltic Sea. Ponto-Caspian invaders are euryhaline species, osmoregulating over a wide salinity range. Owing to the environmental conditions peculiar to the Baltic Sea, it will become the "Sea of Aliens" for newcomers from the Ponto-Caspian region. However, the ecological impact of invading species cannot always be predicted from a knowledge of their biology in their native area. Every new invasion should be considered individually, so further research on biological invasions should be continued.

Conclusion

Ponto-Caspian crustaceans are species with excellent plasticity, which ensures that they have the required broad tolerance to the environmental conditions in the Baltic Sea. The fitness of Ponto-Caspian crustaceans due to their flexibility shows that they have adapted well to their new environment.

References

Bäthe, J, Coring, E., (2011). Biological effects of anthropogenic salt-load on the aquatic Fauna: a synthesis of 17 years of biological survey on the rivers Werra and Weser. *Limnologica* 41: 125 - 133.

Bącela, K., Konopacka, A., (2005). The life history of *Pontogammarus robustoides*, an alien amphipod species in Polish waters. *Journal of Crustacean Biology* 25 (2): 190 - 195.

Berezina, N., (2007). Invasions of alien amphipods (Amphipoda: Gammaridea) in aquatic ecosystems of North-Western Russia: pathways and consequences. *Hydrobiologia* 590: 15 - 29.

Bursey, C. R., Lane, C. E., (1971). Ionic and protein concentration changes during the molt cycle of *Penaeus duorarum. Comp. Biochem. Physiol.* 40a: 155-162.

Bij de Vaate, A., Jażdżewski, K., Ketelaars, H., Gollasch, S., Van der Velde, G., (2002). Geographical patterns in range extensions of macroinvertebrate Ponto-Caspian species in Europe. *Canadian Journal of Fisheries and Aquatic Sciences* 59: 1159-1174.

Braukmann, U., Bohme, D., (2011). Salt pollution of the middle and lower sections of the river Werra (Germany) and its impact on the benthic macroinvertebrates. *Limnologica* 41: 125 - 133.

Casellato, S., La Piana, G., Latella, L., Ruffo, S., (2006). *Dikerogammarus villosus* (Sowinsky, 1894) (Crustacea, Amphipoda, Gammaridea) for the first time in Italy. *Italian Journal of Zoology* 73(1): 97 - 104.

Casellato, S., Masiero, L., La Piana, G., Gigliotti, F., (2008). The alien amphipod crustacean *Dikerogammarus villosus* in Lake Garda (N-Italy): the invasion continues. In: Rabitsch W, Essl F, Klingenstein F (Eds): *Biological Invasions - from Ecology to Conservation.* NEOBIOTA 7: 115 - 122.

Chuberenko, I., Tchepikova, I., (2001). Modelling of man-made contribution to salinity increase into the Vistula Lagoon (Baltic Sea). *Ecological Modelling* 138: 87 - 100.

Chubarenko, B., Margoński, P., (2008). The Vistula Lagoon [in:] Ecology of Baltic coastal waters, Schiewer U. (ed.), *Ecological Studies* 197, Springer-Verlag, Berlin, Heidelberg: 167 - 195.

Chuševe, R., Mastitsky, S. E., Zaiko, A., (2012). First report of endosymbionts in *Dreissena polymorpha* from the brackish Curonian Lagoon, SE Baltic Sea. Oceanologia 54 (4): 701 - 713.

Dedyu, I. I., (1980). Amphipods of the fresh and brackish waters of the South-West USSR, Shtiintsa, Kishinev: 223 pp., (in Russian).

Devin, S., Piscart, C., Beisel, J-N. and Moreteau, J. C., (2004). Life history traits of the invader *Dikerogammarus villosus* (Crustacea: Amphipoda) in the Moselle River, France. International Review of Hydrobiology 89: 21 - 34.

Dick, J. T. A., Platvoet, D., (2000). Invading predatory crustacean *Dikerogammarus villosus* eliminates both native and exotic species. Proc. R. Soc. Lond. B 268: 977-983.

Dick, J. T. A., Platvoet, D., Kelly, D., (2002). Predatory impact of the freshwater invader *Dikerogammarus villosus* (Crustacea: Amphipoda). *Can. J. Fish. Aquat. Sci.* 59: 1078-1084.

Dobrzycka, A., Szaniawska, A., (1995). The effect of salinity on osmoregulation in *Corophium volutator* (Pallas) and *Saduria entomon* (Linnaeus) from the Gulf of Gdańsk. Oceanologia 37: 111-122.

Dobrzycka-Krahel, A., Surowiec, J., (2011). Osmoregulation of *Pontogammarus robustoides* (G.O. Sars, 1894) (Amphipoda) and its distribution in the brackish waters of northern Poland. Crustaceana 84 (14): 1755 - 1767.

Dobrzycka-Krahel, A., Krzak, M., Szaniawska, A., (2014). A laboratory based comparison of osmoregulatory ability at different water temperatures and salinities in the stenothermic isopod *Saduria entomon* (Linnaeus, 1758) and the eurythermic amphipod *Corophium volutator* (Pallas, 1766) from the Baltic Sea. *Marine and Freshwater Behavior and Physiology* 47 (1): 29 - 39.

Dobrzycka-Krahel, A., Graca, B., (2014). Laboratory study of the effect of salinity and ionic composition of water on the mortality and osmoregulation of the gammarid amphipod *Dikerogammarus haemobaphes* (Eichwald, 1841): implications for understanding its invasive distribution pattern. *Marine and Freshwater Behaviour and Physiology* 47 (4): 227 - 238.

Dobrzycka-Krahel, A. & Graca, B., in preparation. *Obesogammarus crassus* (G.O. Sars, 1894) and other Ponto-Caspian gammarids – what affects their invasive success in the environment.

Dobrzycka-Krahel, A., Melzer, M., Majkowski, W., (2015). Range extension of *Dikerogammarus villosus* (Sowinsky, 1894) in Poland (the Baltic Sea basin) and its ability to osmoregulate in different environmental salinities. *Oceanological and Hydrobiological Studies* 44 (3): 294 - 304.

Dobrzycka-Krahel A., Tarała A., Majkowski W., accepted. Estimating fitness: length - weight relationships of alien gammarids in the new expanded brackish environments. *Ecological Indicators.*

Dobrzycka-Krahel, A., Majkowski, W., Melzer, M., under review. Fitting length-weight relationships of Ponto-Caspian gammarids that have overcome the salinity barrier of southern Baltic Sea coastal waters. *Ecological Indicators.*

Fielder, D. S., Bardsley, W. J., Allan, G. L., (2001). Survival and growth of Australian snapper, *Pagurus auratus*, in saline groundwater from inland New South Wales, Australia. *Aquaculture* 201: 73 - 90.

Gallardo, B., Aldridge, D. C., (2013). *Review of the ecological impact and invasion potential of Ponto Caspian invaders in Great Britain.* Cambridge Environmental Consulting: 130 pp.

Gasjunas, I. I., (1972). *Obogaschenie kormovoj bazy ryb vodoemov Litvy akklimatizirovannymi rakoobraznymi Kaspijskogo kompleksa.* [W:] Virbickas J. (ed.), Voprosy razvedenija ryb i rakoobraznykh v vodoemakh Litvy: 57 – 68.

Gherardi, F., (2007). *Biological invaders in inland waters: profiles, distribution and threats.* Springer: 305 pp.

Grabowski, M, Bącela, K, Konopacka, A., (2007 a). How to be an invasive gammarid (Amphipoda: Gammaroidea) – comparison of life history traits. *Hydrobiologia* 590: 75 - 84.

Grabowski, M., Jażdżewski, K. & Konopacka, A., (2007 b). Alien Crustacea in Polish waters - Amphipoda. *Aquatic Invasions* 2 (1): 25 - 38.

Grabowski, M., Bącela, K., Konopacka, A., Jażdżewski, K., (2009). Salinity-related distribution of alien amphipods in rivers provides refugia for native species. *Biological Invasions* 11(9): 2107 - 2117.

Gross, L., (2006). Modeling alien invasions: plasticity may hold the key to prevention. *PLOS Biology* 4 (11): 1888-1890.

Hile, R., (1936). Age and growth of the cisco *Leucichthys artedi* (Le Sueur), in the lakes of the north-eastern highlands, Wisconsin., Bulletin of the Bureau of Fisheries vol. XLVIII no 19: 211-317.

Holland, H. D., (1978). *The Chemistry of the Atmosphere and Oceans.* Wiley, New York: 351pp.

Hørlyck, V., (1973). The osmoregulatory ability in three species of the genus *Idotea* (Isopoda, Crustacea). *Ophelia* 12: 129 - 140.

Ioffe, C. I., (1973). Akklimatizatsionnye fondi bezpozvonotchnykh SSSR, [Acclimatisation stocks of invertebrates in the USSR]. IZV.GosNIORCH 84: 18-68.

Jażdżewski, K., Grabowski, M., (2011). Alien crustaceans along the southern and western Baltic Sea. In: *In the wrong place-alien marine crustaceans: distribution, biology and impacts*. Invading Nature-Springer Series in Invasion Ecology 6, Springer Science + Business Media B. V.: 323-344.

Kefford, BJ. Papas, PJ. Metzeling, L. Nugegoda, D., (2004). Do laboratory salinity tolerances of freshwater animals correspond with their field salinity? *Environ. Pol.* 129: 355 - 362.

Khlebovich, V. V., (1968). Some peculiar features of the hydrochemical regime and the fauna of mesohaline waters. *Marine Biology* 2 (1): 47 - 49.

Khlebovich, V. V., (1974). The Critical Salinity of Biological Processes. *Nauka,* Leningrad: p. 236 (in Russian).

Khlebovich, V. V., (1990). Some physico-chemical and biological phenomena in the salinity gradient. *Limnologica* (Berlin) 20 (1): 5 - 8.

Kinne, O., (1971). Salinity – animals – invertebrates. In: Kinne, O. (Ed.), *Marine Ecology,* vol. 2 (part 2). Wiley Interscience, London: 822 - 995.

Kotta, J., Lauringson, V., Martin, G., Simm, M., Kotta, I., Herkül, K., Ojaveer, H., (2008). Gulf of Riga and Pärnu Bay. U. Schiewer (ed.) *Ecology of Baltic Coastal Waters.* Ecological Studies 197. Springer-Verlag Berlin Heidelberg: 217 - 243.

Konopacka, A., (1998). Nowy dla Polski gatunek kiełża, *Dikerogammarus haemobaphes* (Eichwald, 1841) (Crustacea, Amphipoda) oraz dwa inne rzadkie gatunki skorupiaków obunogich w Wiśle. *Przegląd Zoologiczny,* 3-4: 211 - 218 (in Polish).

Korycińska, M., Królak, E., (2006). The use of various biotic indices for evaluation of water quality in the lowland rivers of Poland (exemplified by the Liwiec River). *Polish J. of Environ. Stud.* 15 (3): 419 - 428.

Malmberg, S. - A., Svansson, A., (1982). *Variations in the physical marine environment in relation to climate.* ICESCM 1982/Gen: 4. Mini symposium.

Mastitsky, S. E., Makarevich, O. A., (2007). Distribution and abundance of Ponto-Caspian amphipods in the Belarusian section of the Dnieper River. *Aquatic Invasions* 2 (1): 39 - 44.

Mc Lusky, D. S., Hagerman, L, Mitchell, P., (1982). Effect of salinity acclimation on 385 osmoregulation in *Crangon crangon* and *Praunus flexuosus*. *Ophelia,* 21 (1): 89 - 100.

Mantel, L. H., Farmer, L. L., (1983). Osmotic and ionic regulation. The Biology of Crustacea. In: Bliss, D. E., Mantel L. H. (eds), *International Anatomy and Physiological Regulation* vol. 5. Academic Press, New York: 53 - 161.

Matthäus, W., Schinke, H., (1994). Mean atmospheric circulation patterns associated with major Baltic inflows. *Deutsche Hydrographische Zeitschrift,* 46: 321-339.

Mordukhai-Boltovskoi, F. D., (1960). *Caspian fauna in the Azov and Black Sea Basin.* Izdatelstvo Akademii Nauk SSSR, Moscow [in Russian].

Motyka, J., Postawa, A., (2000). Influence of contaminated Vistula River water on the groundwater entering the Zakrzówek limestone quarry, Cracow region, Poland. *Environmental Geology* 39 (3-4): 398 - 404.

Launiainen, J., Vihma, T., (1990). Meteorological, ice and water exchange conditions. Second periodic assessment of the state of the marine environment of the Baltic Sea, 1984-1988. *Baltic Sea Environment Proceedings* No. 35 B: 22-23.

Laxson, C. L., McPhedran, K. N., Makarewicz, J. C., Telesh, I. V., MacIsaac, H. J., (2003). Effects of the non-indigenous cladoceran *Cercopagis pengoi* on the lower food web of Lake Ontario. *Freshwater Biology* 48: 2094-2106.

Lee, C. E., Bell, M. A., (1999). Causes and consequences of recent freshwater invasions by saltwater animals. *Perspectives* 14 (7): 284 - 288.

Leppäkoski, E, Olenin, S., (2000). Non-native species and rates of spread: lessons from the brackish Baltic Sea. *Biol. Invasions* 2(2): 151–163.

Lindqvist, O. V., Lathi, E., (1983). *On the sexual dimorphism and condition index in the crayfish, Astacus astacus* L., in Finland. Freshw. Crayfish 5: 3-11.

Luke, A. R., Davis, D. A., Saoud, I. P., Raymond, P. H., (2007). Effects of varying levels of aqueous potasium and magnesium on survival, growth and respiration of the Pacific white shrimp, *Litopenaeus vannamei*, reared in low salinity waters. *Aquaculture:* 461 - 469.

Ojaveer, H., Simm, M., Lankov, A., (2004). Population dynamics and ecological impact of the non-indigenous *Cercopagis pengoi* in the Gulf of Riga (Baltic Sea). *Hydrobiologia* 522 (1–3): 261–269. DOI: 10.1023/B:HYDR.0000029927.91756.41.

Ojaveer, H., Leppäkoski, E., Olenin, S. Q., Ricciardi, A., (2002). *Ecological impact of Ponto-Caspian invaders in the Baltic Sea, European inland waters and the Great Lakes: an inter-ecosystem comparison:* 412-425. Kluwer Academic Publishers, Printed in Netherlands.

Olenin, S., Leppäkoski, E., (1999). Non-native animals in the Baltic Sea: alteration of benthic habitats in coastal inlets and lagoons. *Hydrobiologia* 393: 23-243.

Orlova, M., Golubkov, S., Kalinina, L., Ignatieva, N., (2004). *Dreissena polymorpha* (Bivalvia: Dreissenidae) in the Neva Estuary (eastern Gulf of Finland, Baltic Sea): is it a biofilter or source for pollution? *Mar Pollut Bull.* 49(3): 196-205.

Paavola, M., Olenin, S., Leppäkoski, E., (2005). Are invasive species most successful in habitats of low native species richness across European brackish water seas? *Estuar. Coast. Shelf. Sci.* 64 (4): 738-750.

Peacor, S. D., Allesina, S., Riolo, R. L., Pascual, M., (2006). Phenotypic plasticity opposes species invasions by altering fitness. *PLOS Biology* 4 (11): 2112-2120.

Petruk, A., Stoffler, U., (2011). On the history of chloride concentrations in the River Lippe (Germany) and the impact on the macroinvertebrates. *Limnologica* 41: 143 - 150.

Põllumäe, A., Kotta, J., (2007). Factors describing the distribution of the zooplankton community in the Gulf of Finland in the context of interactions between native and introduced predatory cladocerans. *Oceanologia* 49(2): 277-290.

Pőckl, M., (2009). Success of invasive Ponto-Caspian amphipod *Dikerogammarus villosus* by life history traits and reproductive capacity. *Biological Invasions* 11: 2021 - 2041.

Rajasilta, M., Hänninen, J., Vuorinen, I., (2014). Decreasing salinity improves the feeding conditions of the Baltic herring (*Clupea harengus membras*) during spring in the Bothnian Sea, northern Baltic. *ICES Journal of Marine Science*: 1-5.

Reid, D. F., Orlova, M. I., (2002). Geological and evolutionary underpinning for the success of Ponto-Caspian species invasions in the Baltic Sea and North American Great Lakes. *Can. J. Fish. Aquat. Sci.* 59: 1144 - 1158.

Remane, A., (1934). Die Brackwasserfauna. *Verhandlungen der Deutschen Zoologischen Gesellschaft* 36: 34-74.

Różańska, Z., (1987). Zasolenie. In: *Zasoby, zanieczyszczenia i ochrona wód morskich ze szczególnym uwzględnieniem Bałtyku.* Warszawa. Państwowe Wydawnictwo Naukowe: 30-35.

Surowiec, J., Dobrzycka – Krahel, A., (2008). New data on the non – indigenous gammarids in the Vistula Delta and the Vistula Lagoon. *Oceanologia,* 50 (3): 443 - 447.

Sutccliffe, D. W., (1971). Regulation of water and some ions in gammarids (Amphipoda). J. Exp. Biol. 55: 325 - 344.

Schulz, C-J., (2011). Salinisation of running waters. *Limnologica* 41: 79.

Wolnomiejski, N., Woźniczka, A., (2008). A drastic reduction in abundance of *Dreissena polymorpha* Pall. in the Skoszewska Cove (Szczecin Lagoon, River Odra estuary): effects in the population and habitat. *Ecological Questions* 9/2008: 103 - 111.

Tricarico, E., Mazza, G., Orioli, G., Rossaro, C., Scopini, F. & Gherardi, F., (2010). The killer shrimp, *Dikerogammarus villosus* (Sowinsky, 1894), is spreading in Italy. *Aquatic Invasions* 5 (2): 211 - 214.

Vuarinen, I., Hänninen, J., Viitasalo, M., Helminen, U., Kuosa, H., (1998). Proportion of copepod biomass declines with decreasing salinity in the Baltic Sea. *ICES Journal of Marine Science* 55: 767-774.

Różańska, Z., (1987). Zasolenie. In: *Zasoby, zanieczyszczenia i ochrona wód morskich ze szczególnym uwzględnieniem Bałtyku*. Warszawa. Państwowe Wydawnictwo Naukowe: 30-35.

Zenkevitch, L. A., (1959). The classification of brackish-water basins as exemplified by the seas of the U.S.S.R. *Arch Oceanogr Limnol* 11: 53-61.

In: Crustaceans
Editor: Victor Alvarado
ISBN: 978-1-63484-890-9

Chapter 2

EXTRA AND "INTERMEDIATE" LARVAL STAGES IN DECAPOD CRUSTACEA: A REVIEW OF PHYSIOLOGICAL CAUSES AND ECOLOGICAL IMPLICATIONS, WITH EMPHASIS ON LOBSTERS, *HOMARUS* SPP

Brady K. Quinn*
Department of Biology, University of New Brunswick,
Saint John, NB, Canada

ABSTRACT

Decapod crustacean life cycles include several distinct larval stages, which differ in their morphology, physiology, and behaviour. Generally, under normal conditions larval morphogenesis is synchronized with the moult cycle. In many shrimps and some crabs the number of stages is variable, so extra stages may occur, most often in sub-optimal conditions. This seems to be an adaptive mechanism that lengthens the time available for feeding and growth in preparation for post-larval life in such conditions. In other groups, including homarid lobsters, the number of larval stages appears to be fixed at a species-specific number. Extra stages have been experimentally-induced in these taxa through eyestalk ablation, exposure to toxins, malnutrition, and sub-optimal rearing

* Corresponding author: bk.quinn@unb.ca.

conditions. Resulting extra stages are called "intermediate" stages because they are intermediate in form between particular normally-occurring stages. Intermediate stages result through damage to organs that secrete hormones that regulate the moult cycle, so that moulting occurs in treated animals before the complete morphology of the next stage is prepared. Interestingly, intact larvae in intermediate stages have been observed in the laboratory and field with no apparent cause for their abnormality. This suggests that intermediate stages can result in larvae without damage to moult-inhibiting organs; if so, these stages might actually result in nature from similar adaptive developmental plasticity as that seen in other groups. This has implications to crustacean populations, and fisheries supported by them, because if development of enough larvae is delayed in this way recruitment to adult populations would be delayed. However, most previous research has not considered this aspect of the possible ecological significance of these intermediate stages. In this chapter, reports of extra and intermediate stages of decapod crustacean larvae are reviewed, to better elucidate the nature, occurrence, and implications of intermediate stages. As a case study of an economically-important crustacean group, particular attention is given to lobsters of the genus *Homarus*, which support major fisheries and have been induced to produce intermediate-stage larvae. The time taken by intact and experimentally-treated larvae to complete normal and intermediate larval stages is considered and compared, to determine the state of present knowledge regarding larval development time of intermediate lobster larvae and its implications to lobster populations and fisheries. To facilitate future work, new testable hypotheses are outlined, as are potential methodologies that could be used to in directed studies to investigate the importance of intermediate stages to lobster larval ecology.

Keywords: larvae, crustacean, decapoda, development, plasticity, intermediate stages, extra larval stages, physiology, moulting, lobsters

1. INTRODUCTION

The life cycles of many organisms contain a larval phase that is morphologically and ecologically distinct from the adult. Many benthic marine crustaceans produce pelagic (planktonic and/or nektonic) larvae as a means of increasing dispersal of offspring and creating niche separation between larvae and later life stages, which prevents competitive exclusion of new recruits from populations by established individuals (Anger 2006; Strathmann 1985; Vaughn and Allen 2010). These life cycles present challenges to research because larvae must be studied in a much different manner from adults and their biology differs markedly from them in many respects (Anger 2006; Gore 1985; Vaughn and Allen 2010). However, because aspects of all later life stages (condition, growth, abundances, population connectivity, etc.) depend on what happens during the larval phase studies of larval biology are essential to understanding dynamics of these species (Strathmann 1985).

Given the above, it is clear that anything which alters the length or nature of larval development can have far-reaching impacts on larval and post-larval survival and recruitment to populations (Morgan 1990; Strathmann 1985) and is thus an essential component of modeling life history in these species. Delayed completion of pelagic larval development can be very disadvantageous because it lengthens the time larvae are exposed to pelagic predators (Morgan 1990; Vaughn and Allen 2010) and therefore increases the likelihood of larval mortality; this also increases the probability that larvae will be advected away from their natal population (reducing local recruitment; Quinn 2014) or into areas lacking suitable habitat for settlement to the benthos (Incze and Naimie 2000). Estimating the impacts of abiotic factors, for example future climate change, on larval survival and crustacean populations thus requires knowledge of the specific developmental sequence of larvae for particular species.

Traditionally, each species was thought to go through a single, fixed developmental sequence with minimal variation in the duration and morphology of each phase or stage of this sequence among individuals of the same species (Anger 2001). Extra stages (= stages not observed among the majority of larvae and/or in the original descriptions of larvae of a given species) were occasionally observed among these taxa, but were believed to represent rare anomalies, perhaps related to larval rearing conditions (Anger 2001; Gore 1985; Gurney 1942). However, recent research has revealed that in many if not nearly all crustacean species the number and types of stages occurring within development sequences can vary considerably among conspecific individuals (Anger 2001; Gore 1985; Knowlton 1974; Williamson 1982). Such developmental variability or plasticity is now particularly well-known to occur within certain taxonomic groups and is seen as a natural, adaptive response to suboptimal conditions (Anger 2001; Oliphant et al. 2013). In other groups developmental sequences are still thought to be fixed, but extra developmental stages have been induced to occur in these, often with intermediate morphology between certain normally-occurring stages (Charmantier and Aiken 1987; Costlow 1966a; Gore 1985). Studies creating these “intermediate” stages helped to elucidate aspects of developmental and moult cycle control, but intermediate stages themselves were thought to have little real-world ecological relevance (Anger 2001; Gore 1985; Webster 2015). The possibility remains, though, that these could occur in nature and represent an important and thus far unexplored component of development in these species.

In this chapter, the occurrence of extra larval stages among crustaceans is reviewed. The focus is primarily on decapods because this group is particularly well-studied and contains many species that support major human fisheries and are thus of considerable socio-economic importance. Where information is available, reference is also made to other groups within the Crustacea and Arthropoda. Larval development in the Crustacea in general is first reviewed, including some discussion of terminology and the physiological processes involved in coordinating development and the closely-related moulting cycle. Occurrence of extra larval stages in various taxonomic groups is then described, including both those known to result from natural causes (i.e., developmental plasticity) and those induced by experimental manipulations (hereafter termed "intermediate" stages, although note that many of the "natural" extra stages also show intermediate morphology). The causes and ecological implications of extra larval stages are then discussed. Induced intermediates are compared to extra stages occurring in naturally-plastic developmental sequences. The argument is made that intermediates could represent heretofore underappreciated components of normal larval development, in which case they have important ecological implications and require further study. Finally, larval development of lobsters in the genus *Homarus* is reviewed as a detailed case study of an important decapod genus in which intermediate stages have been induced and could strongly impact predictions made regarding species biology and fisheries ecology.

2. General Overview of Crustacean Larval Development

2.1. Larval Phases and Stages

Excluding highly-specialized groups (e.g., obligate parasites), the total larval period in decapod crustaceans can be subdivided into three main phases that always occur in the same order. These are (from least to most developmentally advanced) the nauplius, zoea, and decapodid phases (Martin et al. 2014; Williamson 1982). All crustaceans go through some form of the nauplius phase, but groups that diverged earliest do not go through the other two phases as these presumably had not yet evolved when they diverged (Williamson 1982); many later-diverged non-decapod groups go through nauplius and zoea phases, but not the decapodid one, while the more highly-derived Decapoda have all three phases (Gore 1985). Numerous morphological characters distinguish these three phases, most notably that the posteriormost (e.g., abdominal) segments and appendages are initially poorly developed but then become more elaborated as development proceeds within and especially between these phases (Gore 1985; Williamson 1982). The most straightforward distinction among these phase is behavioural, however, specifically in terms of which appendages are used for locomotion: the nauplius mainly uses cephalic (head) appendages (antennae or maxillipeds) to move, the zoea uses thoracic appendages (exopodites on the pereiopods), and the decapodid uses abdominal ones (pleopods) (Williamson 1982); post-larval juveniles typically walk or swim with the pereiopods, though some also continue to swim using their pleopods and/or abdominal flexion (Charmantier and Aiken 1987; Felder et al. 1985). It should be noted that the decapodid phase is called by various terms in the literature, including those specific to certain groups (megalopa, glaucothoea, etc.) or the contradictory term "postlarva" (properly a post-larval phase is not a larva, but the decapodid phase is; Anger 2001; Gore 1985; Martin et al. 2014), without much consistency. To be consistent with current standards (e.g., Martin et al. 2014) and avoid confusion across taxa, stages within the larval phase following the zoeae and preceding the juvenile (= truly post-larval stages) stages are always called decapodid throughout this chapter.

Current consensus is that crustaceans do exhibit metamorphosis from larvae to adult forms, as is well-known in insects, but that the metamorphosis tends to be more gradual and occurs over multiple moults (Passano 1961). A complimentary view is that crustaceans undergo multiple metamorphoses, with each falling between one larval phase and the next in addition to between the last larval phase and the juvenile (Charmantier et al. 1991). In taxa with relatively abbreviated development, one or more of these phases can be "skipped," meaning that it does not occur among free-living larvae but is instead completed within the egg prior to hatching (Gore 1985); in decapod crustaceans, for instance, the naupliar phase is completed entirely within the egg and larvae hatch as zoeae (Anger 2001). In these cases, developmental changes from one larval phase to the next tend to be more dramatic, resulting in a "more metamorphic" developmental pattern (Rötzer and Haug 2015).

Each of the phases described above can and usually does consist of multiple larval stages. Earlier stages within each phase are the least advanced (i.e., possess features of that phase only) and then subsequent stages gradually acquire and develop features of the next phase (Gore 1985; Williamson 1982); for example, the earliest zoeal stages of the lobster *Homarus americanus* lack pleopods and uropods, which are decapodid features, but these appear in later zoeal stages (Charmantier et al. 1991; Ennis 1995). There is thus a clear developmental progression from one stage to the next that can be observed and used to distinguish stages. Each stage generally begins and ends with a moult and is distinct from other stages in its morphology, behaviour, and physiology. However, in some groups multiple moults can occur within the same developmental stage. This often occurs during the highly-specialized zoeal stages (called phyllosomas) of spiny and slipper lobsters (Decapoda: Infraorder Achelata) (Phillips and Satry 1980), as well as among other groups with relatively long pelagic larval phases (e.g., Gore 1985; Gurney 1936; Gurney and Lebour 1941; Ngoc-Ho 1976; Williamson 1976). While larvae can increase in size (grow) following such moults, they do not progress in development and thus these moults do not demarcate separate *stages*; instead the same stage is repeated, so the term *instar* should be applied (Phillips and Sastry 1980; Martin et al. 2014).

2.2. Control of Moulting and Developmental Cycles

Larval development and growth in the Crustacea are tightly tied to the moult cycle because new or changed body structures cannot be formed, nor can size increase very much, through the hard cuticle forming the crustacean exoskeleton (Anger 2001; Costlow 1965; Gore 1985). In the "classical" model of moult regulation in the Crustacea whether and when moulting occurs is determined by the relative titres of two neurohormones in the hemolymph: moult inhibiting hormone (MIH) and ecdysteroids, mainly β-ecdysone (Webster 2015). MIH is produced primarily by the X-organ-sinus-gland complex located in the eyestalks and prevents moulting, while ecdysteroids are produced by the Y-organ located around the base of the mandibles and stimulate moulting (Freeman and Costlow 1980; Wesbster 2015). Early in a single moult cycle (e.g., intermoult period after completion of a previous moult) levels of ecdysteroids are very low while MIH levels are relatively high; as the intermoult period advances, MIH titres begin to fall and ecdysteroid levels increase (Webster 2015). This usually occurs around moult stage D_0 (Drach 1949) and is the point at which an animal becomes committed to an eventual moult (Anger 2001; Gore 1985). Recently other hormones (methyl farnesoate (MF), crustacean hyperglycemic hormone (CHH), etc.) and organs (e.g., mandibular organs) have been identified as contributing to moult cycle control as well; for more, see Webster (2015). There is feedback between the moult cycle components and other parts of the crustacean neuroendocrine system, which can allow moult to be delayed (Nelson 1991); for example if energy reserves are too low as a result of low food levels or stressful environmental salinities or temperatures moult can be delayed until sufficient energy reserves have been accumulated or conditions improve (Anger 2001; Gore 1985). Feedback mechanisms are unclear, but likely involve additional neurohormones that act as messengers and induce changes in titres of MIH and ecdysteroids (Nelson 1991; Webster 2015).

Development and growth are controlled by separate physiological mechanisms from moulting and each other (Costlow 1962; Freeman and Costlow 1980). Though both appear discontinuous because they can only manifest visually after a moult, they are actually continuous processes which continue throughout the intermoult period (Gore 1985; Hartnoll 1982). Cell division and apoptosis producing new appendages and setae occur throughout the moult cycle in larvae and are regulated by their own neurohormones (Anger 2001), though the nature and identities of these regulatory agents of development and morphogenesis are not yet well-known (Webster 2015). One potentially-important neurohormone to larval development is MF, which is actually the precursor to an insect hormone called juvenile hormone (JH) (Webster 2015). JH is produced by larval insects and prevents the final, metamorphic moult from larva or pupa to adult (Esperk et al. 2007); metamorphosis thus occurs once JH production has ceased. In cases where metamorphosis is delayed, for example due to poor condition following low food levels (Safranek and Williams 1984a, b), JH production continues for longer than usual. Though crustaceans do not produce JH or any confirmed equivalent, insect JH has been shown to delay metamorphic moults in crustaceans (e.g., Charmantier et al. 1988; Tighe-Ford 1977). Crustacean MF does seem to delay crustacean metamorphosis, so it may play a similar role to JH in insects, but this has yet not been conclusively determined (Webster 2015).

Whatever the components of development cycle regulation in crustaceans, there appears to be some interaction between them and the moult cycle (Nelson 1991). In many species (see review below in section 3, especially sections 3.2.3 and 3.2.4) larvae that are undernourished and/or exposed to stressful environmental conditions will moult sooner, pass through more stages, and complete development more slowly overall than normal (e.g., Oliphant et al. 2013); alternatively, under some conditions stages can be skipped (Gore 1985; Williamson 1982). There are several points during the moult and development cycles that determine the course of future events in these cycles, for instance based on nutritional condition at that point (Anger 2001; Gore 1985). For example, there seems to be a critical point in the development of decapod crustaceans, specifically moult stage D_0 during the second zoeal stage, at which whether and by how much metamorphosis will be delayed is determined (Chamantier et al. 1984a; Costlow 1963a, b). Larvae in these species can thus select their development sequence to optimize their chances of successfully completing the larval phase and recruiting to adult populations.

3. A Review of Extra and Intermediate Larval Stages in the Crustacea

3.1. General Terminology

Larval development and subsequent metamorphoses and settlement in crustaceans can be delayed through two main mechanisms: (1) the addition of extra moults, or instars, without development progressing between moults, a process termed mark-time moulting (Gore 1985), or (2) addition of one or more extra larval stages within the developmental sequence, either within the zoeal stages (= intercalated stages; Gore 1985) or between the final ("ultimate") zoeal phase and the decapodid(s) (= terminally-additive staging; Gore 1985). As the former (mark-time moulting) results in extra instars rather than additional stages it will not be dealt with in the present chapter beyond the present section. It is discussed here because there appears to have been some confusion in previous studies and reviews of larval development between instars and stages. For example, extra stages with distinct morphology, behaviour, etc. have been described as instars (e.g., Oliphant et al. 2013; Williamson 1982) and extra instars with the same characters as normal stages have been described as stages (Williamson 1982); the two terms are often erroneously used interchangeably. However, the distinction matters because added instars merely repeat the same stage without changes to larval characters (save perhaps size), while extra stages result in ecologically-distinct larvae (Martin et al. 2014). It should be noted, though, that similar factors to those that cause extra stages also result in mark-time moulting (e.g., suboptimal conditions; Gore 1985; Gurney 1936; Gurney and Lebour 1941; Williamson 1976) and the two mechanisms of developmental delay often co-occur within the same taxa (e.g., hippolytid shrimps (Decapoda: Infraorder Caridea: Family Hippolytidae); Williamson 1982), so perhaps the two represent different points on a spectrum, with one or the other being more favourable and selected for under particular circumstances; this may be worth further investigation.

There has also been some apparent confusion concerning whether development is delayed by addition of stages within a species versus among species (or higher taxa). Both perspectives are often discussed within the same reviews (e.g., Gore 1985; Williamson 1982), but they refer to different things. For example, the majority of true crabs (Decapoda: Infraorder Brachyura) go through 5 zoeal stages, but a few species regularly pass through 6 or more zoeae, so the 6th-plus zoeae in such species are "extra" in relation to other Brachyuran species and the number of stages within the Brachyura can be said to be variable (Williamson 1982). However, these stages are not extra in relation to such species themselves as they are part of their usual developmental sequences. Development within a species is variable and contains truly extra stages if some conspecific individuals pass through more larval stages than others. The review by Williamson (1982, Table I, p. 54), for instance, in which usual stage numbers of various taxa were listed along with known variations, has been cited as demonstrating variable stage numbers within species by later studies (e.g., Gore 1985) but this was incorrect. While variation in development among species or higher taxa is of interest from an evolutionary perspective, it is of less importance to ecology or fisheries and is thus not dealt with in this chapter; instead intraspecific differences in development are the focus. The interested reader is referred to Gore (1985) and Williamson (1982) for further information on inter-taxon variability.

Reports of "abnormal" crustacean larvae, meaning those belonging to a particular known species but displaying different or intermediate morphology from that shown by known larval stages, have been made since very early in the history of carcinology, for example by Faxon (1879), Brooks and Herrick (1892), and Lebour (1928, 1930). Later studies (e.g., Carli 1978; Dakin 1938; Knowlton 1974; Le Roux 1966; Sandifer 1973a, b; Templeman 1936; Yatsuzuka 1962) observed similar unusual larvae among a variety of taxa, which lead to the realization that the types of larval stages occurring during development could vary within a species. Specifically, conspecific individuals sometimes undergo very different developmental sequences from one another, usually as a result of different conditions experienced during development. This phenomenon has been termed poecilogony and occurs within many species, including but not limited to the Crustacea (Gore 1985). Early reports were made on shrimps (Caridea) in the Families Atyidae (Gurney 1942), Alpheidae (Brooks and Herrick 1892; Knowlton 1970, 1973), Hippolytidae (Le Roux 1963), and Palaemonidae (Shokita 1977), as well as mud lobsters (Infraorder Thalassinidea) in the Families Upogebiidae (Webb 1919; Shenoy 1967) and Laomediidae (Goy and Provenzano 1978) and some crabs (Anomura: Fageti et al. 1971; Brachyura: Williamson 1965). Many of these early larval studies within the Crustacea that reported poecilogony focused on the morphological differences among larvae from different sequences and imply that the different larval forms observed represent different morphs or even incipient species. Alternatively, some investigators interpreted poecilogonous larvae as "substages" of normally-occurring stages (e.g., Boyd and Johnson 1963; Fageti et al. 1971; Gore 1985). However, these are not the only or best interpretations possible for these observations. The different larval types reported clearly showed more or less advanced stages of development in terms of appendage and carapace morphology, setation, etc., which rather implies larvae belonging to more or less delayed developmental series. In the present chapter, reports of larvae different from the usual type (or showing multiple "normal" types) as a result of poecilogony are thus considered as representing extra larval stages belonging to delayed development series. However, it would be worthwhile for these earlier studies to be followed-up on from this perspective to confirm the true nature of the different larval forms observed.

3.2. Extra Stages Occurring Naturally or From Unknown Causes

3.2.1. Extra Stages Occurring among the Decapoda

Studies that reported extra larval stages in Crustacea were obtained for review primarily from the reviews by Gore (1985), Knowlton (1974), Williamson (1982), and Anger (2001), although additional studies were obtained through searching citations within the papers obtained from these reviews and Web of Science and Google Scholar searches for more recent literature. This review was by no means comprehensive, but was extensive and allowed considerable information to be obtained. Table 1 lists examples of decapod crustacean species (all within the Suborder Pleocyemata) in which extra larval stages (either "natural" or induced stages with intermediate morphology) have been reported, obtained from this literature review. Extra larval stages were reported in several Anomuran crabs (especially the squat lobsters (Family Galatheidae) and porcellanids (Porecellanidae)), the true crabs (Brachyura, especially the grapsid crabs in Superfamily Grapsoidea), many shrimps (Caridea: Families Alpheidae, Hippolytidae, Crangonidae, Palaemonidae, and Pandalidae), and the mud lobsters (Infraorder Thalassinidea) (Table 1). These were also reported less-frequently in other taxa, including *Homarus* spp. clawed lobsters (Infraorder Astacidea: Family Nephropidae) and other families in the Anomura and Brachyura (Table 1). In general, extra-stage larvae have morphological and behavioural characteristics intermediate between those of particular normally-occurring stages, for example between zoeae II and III, the final zoea and the decapodid stages, or even between decapodid and juvenile stages (Gore 1985; Table 1).

In addition to the specific examples listed in Table 1, other more general examples can be cited. Within the decapod Infraorder Pleocyemata, variable development with extra stages and/or poecilogony has been widely reported within the Thalassinidea (Sandifer 1973a, b), the polychelid lobsters (Infraorder Polychelida: Superfamily Eryonoidea) (Williamson 1982), and the Caridea (Knowlton 1974; Le Roux 1963; Makarov 1968; Montú et al. 1990), especially the Panaeidea (Dakin 1938) and Pandalidea (Rothlisberg 1980). This also occurs, although slightly less frequently, among the Anomura (Le Roux 1966; Knowlton 1974) and Brachyura (Knowlton 1974; Montú et al. 1990; Yatsuzuka 1962). Developmental variability has also been reported in decapods outside of the Pleocyemata, within the Suborder Dendrobranchiata, Superfamilies Segestoidea and Penaeoidea (Knowlton 1974; Williamson 1982).

An unusual larval form is occasionally found among the spiny and slipper lobsters (Pleocyemata: Achelata) which has morphology intermediate between a nauplius and phyllosoma (zoea) and quickly moults into a normal early-stage phyllosoma (Phillips and Sastry 1980). The status of this type of larva, termed a "naupliosoma," is unclear: while it may represent an extra, intermediate stage like those seen among zoeae and decapodids in other taxa (Table 1) it might alternatively be a prematurely hatched larva belonging to one of the normally-occurring prezoeal stages that typically elapse within the egg (Phillips and Sastry 1980). Thus, it is uncertain whether extra stages occur among the Achelata, although this group certainly has very flexible development in terms of instar number (see section 2.2, above).

In general, variable numbers of larval stages is very common in the "lower" Decapoda, among groups commonly referred to as "shrimps" or "shrimp-like" species (e.g., Caridea; Montú et al. 1990). There seem to be certain stages or phases within these groups that are obligatory, and thus all individuals must pass through them during development, while other stages are non-obligatory and can be added if conditions require them. Usually the earliest zoeal phases and perhaps the final or ultimate zoea are obligatory, while the number and types of later zoeal stages can vary depending on environmental conditions (Gore 1985; Le Roux 1963; Makarov 1968). Species within these groups tend to pass through more larval stages with less developmental advancement through each stage than do those in the "higher" Decapoda (crabs, lobsters, etc.), which tend to show more abbreviated (Gore 1985) and metamorphic (Rötzer and Haug 2015) development. Further, among those higher decapod groups (e.g., Brachyura) that have been observed to show variable development, additional stages tend to be more common among those with longer developmental series (i.e., more stages) than among those with more abbreviated development (Gore 1985; Montú et al. 1990). This suggests that the number of stages and rate of developmental advancement among them is related to the capacity for developmental plasticity and insertion of extra stages. Perhaps the more ancestral condition for larval development among the Decapoda was a longer, more variable series of larval stages, which is better able to deal with environmental variability (e.g., development adjusted in relation to suboptimal conditions, see sections 3.2.3, 3.2.4), while the more-derived condition is an abbreviated, less variable (i.e., canalized) series of stages that is well-suited to particular, specialized conditions. However, variability has been reported even among species with very short larval series (e.g., two zoeae and one decapodid, Gore 1985; Table 1), so there does appear to still be some capacity for developmental plasticity among higher decapod groups.

3.2.2. Extra Stages occurring Outside the Decapoda

Developmental plasticity has been reported among some copepods (Class Maxillopoda: Subclass Copepoda; Hartnoll 1982) and barnacles (Maxillopoda: Subclass Thecostraca: Infraclass Cirripedia; Williamson 1982). Recently, intriguing reports have been made of the occurrence of extra larval stages of some decapods and also mantis shrimps (Class Malacostraca: Subclass Hoplocarida: Order Stomatopoda) among Mesozoic crustacean fossil deposits, with morphology intermediate between zoeal and decapodid stages occurring in the same deposits and presumably of the same species (Haug C et al. 2009;

Haug JT et al. 2009, 2011); this shows that variable development has been present in the Crustacea for a substantial portion of their existence. Variable numbers of larval stages have been reported frequently among the krill and other euphausiids (Class Malacostraca: Superorder Eucarida: Order Euphausiacea) (Knowlton 1974; Makarov and Maslennikov 1981; Williamson 1982). Gurney (1942) noted that the number of "furcilia" stages (= zoeae) among the Euphausiacea is highly variable, and Fraser (1936) went so far as to suggest that no fixed larval stages occur within this group, but rather development proceeds gradually over a series of moults such that the size and timing at which different points in development (e.g., appendages additions) occur varies among individuals (Fraser 1936; Gore 1985). Perhaps in groups originating earlier than the Decapoda the tight coordination of the moult and developmental cycles (see section 2.2) seen in the higher decapods had not yet evolved, allowing for more variable developmental series (Anger 2001).

Variation in the number of larval stages is also well known to occur among non-crustacean arthropods, especially insects (Esperk et al. 2007; Oliphant et al. 2013). In general, this occurs in the insect Orders Lepidoptera (moths, etc., but excluding the butterflies, Suborder Papilionoidea), Blattodea (roaches), and Hymenoptera (Suborder Symphyta (sawflies) only) (Esperk et al. 2007). One often-cited example is the tobacco hornworm *Manduca sexta* (Lepidoptera: Family Sphingidae), which can delay metamorphosis and go through extra larval and even pupal stages under certain conditions, such as when food supply is limited (Safranek and Williams 1984a, b). As control of the moult and development cycles is very similar among crustaceans and insects (see section 2.2; Webster 2015), it makes sense that similar developmental plasticity occurs among these two separate but related arthropod groups.

3.2.3. Postulated Causes of Extra Larval Stages

Variation in the number of larval stages within arthropod species has previously been linked to various factors, including chemicals present in the rearing environment (antibiotics, pollutants, etc.), density, heredity, humidity, food quality and/or quantity, photoperiod, salinity, sex, and temperature (Esperk et al. 2007; Knowlton 1974; Oliphant et al. 2013; Rochanaburanan and Williamson 1976; Sandifer and Smith 1979; Table 1). In general, the minimal number of larval stages occurs at conditions optimal for larval survival, while extra larval stages are inserted into developmental sequences as a result of larvae experiencing suboptimal conditions during development (Anger 2001, 2006; Knowlton 1974; Gore 1985; Rochanaburanan and

Williamson 1976; Sandifer and Smith 1979; Williamson 1982; Table 1). For example, more larval stages occur at stressfully low salinities among marine species (Christiansen et al. 1977; Linck 1995), and both extremely low (Oliphant et al. 2013) and high (Knowlton 1974) temperatures can result in more larval stages when these exceed the thermal regime to which a species is usually exposed. In this chapter, the above environmental factors were all found to be linked to occurrence of extra larval stages (Table 1), in addition to some others. Interspecific hybridization, acidified seawater (Astacidea: Nephropidae; Hall and Boweden 2012; Agnalt et al. 2015), and lack of substrate suitable for settlement (Caridea: Hippolytidae; K.N. Sankolli, pers. comm. cited in Williamson 1982, p. 81) were also reported to result in extra larval stages (Table 1), and the latter of these clearly represents delay of metamorphosis under suboptimal conditions.

Differences in the occurrence of extra larval stages, even among larvae exposed to stressful conditions, have been reported among larvae hatching from different broods (Boyd and Johnson 1963; Charmantier and Aiken 1987; Linck 1995; Templeman 1936; Williamson 1965; Table 1). Such differences could result from genetic differences among individuals, inherited from their parents, which predispose larvae to greater or lesser developmental plasticity. Alternatively, differences could arise from maternal effects such as differential investment in yolk by females depending on their condition when spawning, which impacts larval condition and development rate. Geographic origin of females producing larvae also impacts the number of larval stages that occur (Criales and Anger 1986), meaning that within the same species particular (sub-)populations might exhibit different developmental series, perhaps as a form of local adaptation to conditions in certain locations (e.g., near the range limits of a species).

In addition, extra larval stages have been observed without any clear cause among larvae reared in the laboratory and sampled from nature in many species (e.g., Costlow 1965; Goy and Provenzano 1978; Knowlton 1974; Nicols and Keney 1963; Pike and Williamson 1961; Tully and Ó Céidigh 1987; see many other examples listed in Table 1). This suggests that such larval stages may not only occur under stressful conditions or as artefacts of laboratory or hatchery rearing, but actually do occur in nature and could simply occur as part of the regular developmental series of certain larvae, for example as a result of genetic differences among individuals. This might also suggest that other factors mediating occurrence of these stages are at play that have not yet been identified. If so, future research efforts should aim to identify these factors, as well as to better quantify in general how often extra stages occur in nature.

The extent to which larval development is variable versus fixed appears to vary across a spectrum within the Crustacea, from those species with highly-variable development and stages ultimately determined by environmental conditions (e.g., euphausiids) to those with genetically-fixed, canalized developmental pathways (e.g., those with direct development, like crayfish) (Anger 2001; Higgins and Rankin 1996). Importantly, while many higher Decapoda were traditionally thought to have fixed developmental pathways, recent evidence suggests that truly fixed development is rare among the Crustacea (Anger 2001, 2006; this chapter). Ultimately, the range of developmental pathways available to an individual larva is determined by its genetic makeup, while which pathway it actually takes during development is determined within this range by external factors like temperature and food levels (Anger 2001; Gore 1985; Higgins and Rankin 1996). These factors interact with the crustacean neuroendocrine system (see section 2.2) and determine the titres of neurohormones controlling the moulting cycle (e.g., MIH, β-ecdysone) and likely also development and metamorphosis (JH-like factors, MF, etc.) (Nelson 1991; Higgins and Rankin 1996).

3.2.4. Ecological Implications of Extra Stages and Developmental Plasticity

Suboptimal conditions tend to limit growth of larvae from stage to stage (or between moults), such that at the end of the larval phase a larva developing under such conditions will moult into a much smaller juvenile than would one that experienced optimal ones (Hartnoll 1982; Oliphant et al. 2013). Post-larval survival in many species, especially those in which juveniles recruit to shallow benthic or intertidal environments, is highly-size dependent (Taylor 2003; Lawton and Lavalli 1995). This is because larger and/or faster-growing juveniles are vulnerable to fewer predators over a shorter time period than smaller and/or slower-growing ones (Taylor 2003; Lawton and Lavalli 1995). If development in a given species is fixed, then larvae developing under suboptimal conditions are less likely to survive the juvenile phase and ultimately reproduce as adults. However, if development is variable this limitation can be overcome. Larvae that delay metamorphosis to the juvenile phase and instead pass through additional larval stages are able to "catch up" under suboptimal conditions and ultimately reach a comparable size to those that developed through fewer stages under optimal conditions (Broad 1957; Hartnoll 1982; Oliphant et al. 2013). This occurs not only among crustaceans, but also insects that can pass through extra larval stages. For example, tobacco hornworm larvae will only metamorphose if they have reached or exceeded a critical weight, but if they have not (e.g., as a result of being malnourished during rearing) they will delay metamorphosis and go through extra larval stages (Safranek and Williams 1984 a, b). The evidence available thus far does not support the existence of a similar critical weight in any crustaceans (e.g., Oliphant et al. 2013), but there is a strong link between diet and occurrence of extra larval stages in Crustacea (see examples in Table 1) so some form of feedback mechanism must be at work to alter developmental pathways based on larval condition. Importantly, larvae are also quite vulnerable to numerous agents of mortality, including predators (Morgan 1990; Vaughn and Allen 2010), so there is a cost to extending larval duration to improve post-larval survival. Presumably whether or not a larva extends its development is decided based on numerous neuroendocrine feedbacks within the individual larva, as well as over evolutionary time for larvae of a given species based on the conditions larvae and juveniles of that species are more likely to face.

Ecology and demography within species with plastic and variable development depends strongly on this aspect of their life history. As such, studies examining these species and attempting to estimate their ecology and demography must account for the fact that the nature and length of the larval phase varies among individuals (Anger 2006); otherwise estimates used to inform conservation, fisheries management, and so will be in error, for instance by underestimating larval duration at extreme temperatures because the addition of extra stages is not modeled. Ideally, the researcher studying such species would know the range of possible developmental sequences, how these differ in terms of larval development time, survival, etc., with what frequency the different sequences occur, and what factors (e.g., temperature and salinity) impact which sequences occur and in what proportion of larvae. In the majority of cases such quantitative data is lacking so most estimates are made based on the "normal" (= most frequent?) developmental sequence. While this approach may be sufficient in much of a species' range it will make important errors in many portions of the range and will not be able to estimate effects of environmental changes to suboptimal conditions, for instance with future climate change. Therefore more study of extra larval stages and developmental plasticity is needed in the future to enhance the quality and predictive power of carcinology in general.

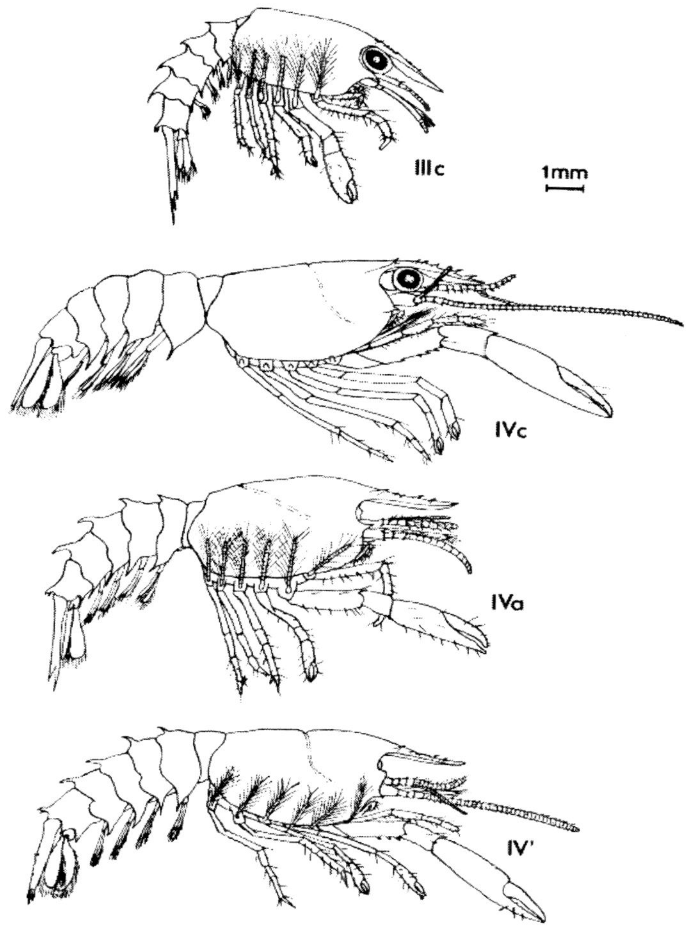

Figure 1. Examples of normal (c = control, upper two larvae) and extra (lower two larvae) larval stages of a decapod crustacean, the American lobster *Homarus americanus*. Extra stages designated IVa (less developmentally-advanced intermediate) and IV′ (more advanced) were produced by bilateral eyestalk ablation (and thus lack eyes) and were intermediate in form between the final zoeal stage (III) from which they moulted and the normal decapodid (stage IV). Note retention of larval characters including short antennal flagellum, dorsal abdominal spines, long setae-bearing exopodites on pereiopods, and achelate third pereiopods alongside decapodid morphology in intermediate-stage larvae; see also Fig. 2. Figure from Charmantier and Aiken (1987), © The Crustacean Society, 1987, published by Brill NV, Leiden, reproduced with permission from the publisher.

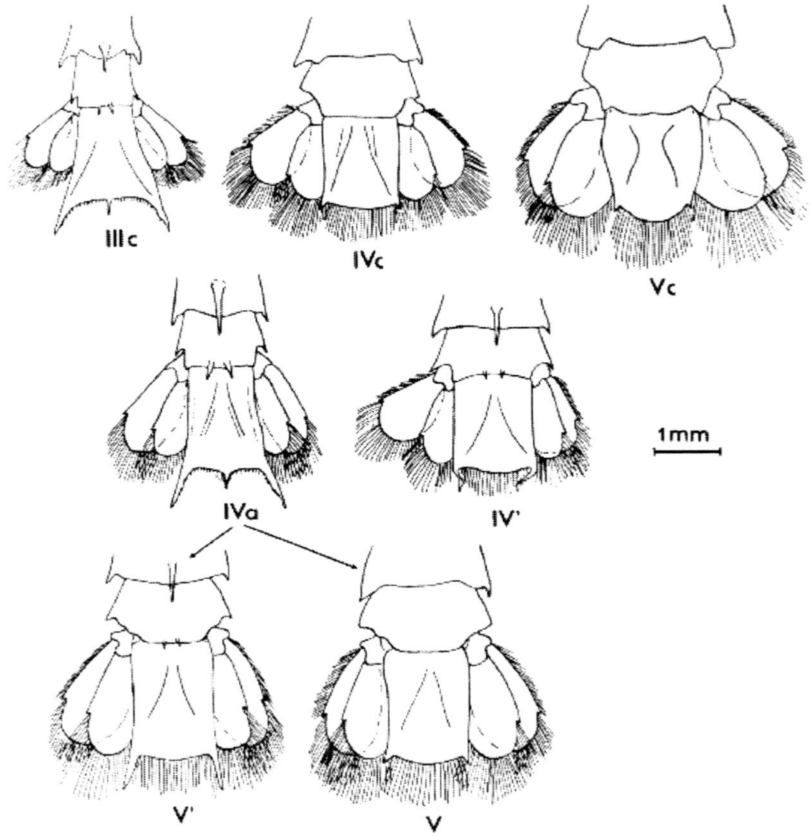

Figure 2. Further details of telson morphology distinguishing normal (c = control) *Homarus americanus* zoea III, decapodid (stage IV), and juvenile (stage V) from intermediate stages induced by eyestalk ablation (IVa, IV′, and V′). Note retention of larval characters, including dorsal abdominal spines, long posterior spines on telson (medial in IVa, posteriolateral in all intermediates) and rectangular rather than rounded shape of telson in intermediates; see Fig. 1 for more. Figure from Charmantier and Aiken (1987), © The Crustacean Society, 1987, published by Brill NV, Leiden, reproduced with permission from the publisher.

3.3. Induced Intermediate Stages

3.3.1. Decapod Examples

The previous section (3.2) outlined occurrence of extra larval stages among crustaceans known to have more plastic development in response to suboptimal conditions that might be encountered in nature. In addition to these, extra "intermediate" stages have also been induced to form among species thought to have more fixed developmental sequences by experimental treatment of larvae, either through surgical injury of larvae or exposure to abnormal pollutants (Table 1). As with "natural" extra stages outlined earlier, these induced extra stages were inserted between normally-occurring ones, especially between the last zoea and decapodid, and showed morphology and behaviour intermediate between these (Charmantier and Aiken 1987; Figure 1, 2, Table 1). Such intermediate stages have been produced by bilateral eyestalk ablation, in which both eyes are removed and the X-organ-sinus-gland complex within the eyestalks is heavily damaged, among porcellanid crabs (Anomura: Family Porcellanidae), homarid lobsters (Astacidea: Family Nephropidae; Figure 1, 2), crabs (Brachyura) in Families Sesarmidae and Panapeidae, and shrimps (Caridea) in Families Hippolytidae and Palaemonidae (Table 1). Anthropogenic chemicals were also found to induce intermediate larval stages among the Xanthoidea (insecticides), Palaemonidae (mercury), Nephropidae (crude oil, toxic potash brine, alkylphenols (known endocrine disruptors), and injection with insect JH) and Panapeidae (JH analogues) (Table 1).

3.3.2. Examples Beyond the Decapod Crustacea

Intermediate larval stages have also been induced in arthropod taxa outside of the decapod Crustacea. Most of these cases were induced by injecting larvae with JH, a neurohormone produced in insect larvae until they reach the point of metamorphosis which prevents the final metamorphic moult (see 3.2). Injection of insect larvae or pupae with JH results in the delay of the next phase and production of an extra pre-adult stage (e.g., tobacco hornworm: Safranek and Williams 1984a, b). Many insecticides are actually JH analogues, which work by delaying metamorphosis indefinitely until larvae or pupae die without reproducing (Esperk et al. 2007). Barnacle (Crustacea: Cirripedia) larvae normally pass through several naupliar stages, which are then followed by a highly-modified cyprid stage that settles and metamorphoses to a sessile juvenile (Williamson 1982). Tighe-Ford (1977) found that exposing barnacle larvae to JH analogues resulted in the last nauplius phase moulting to an abnormal larval stage with morphology intermediate between a nauplius and cyprid, or in production of extra cyprid phases. These nauplius-cyprid intermediates and extra cyprids did not settle and metamorphose to the juvenile stage, and thus the completion of the larval phase of barnacles was delayed by production of these intermediate stages.

3.3.3. Causes of Intermediate Stages

Bilateral eyestalk ablation injures the X-organ-sinus-gland complex, which is one of the major sources of moult inhibiting hormone (MIH) in the crustacean neuroendocrine system. The relative titres of MIH and ecdysteroids (e.g., β-ecdysone, which is released from the Y-organ and promotes moulting and ecdysis) determine the rate at which the moult cycle progresses and ultimately whether and when moulting occurs (see section 2.2; Webster 2015). Ablation thus rapidly reduces titres of MIH, resulting in an early moult and subsequently more rapid moulting rates (Costlow 1965; Le Roux 1979; Webster 2015). Separate physiological cycles control growth, development, reproductive maturation, etc., so the developmental cycles, which control when appendages are added, elaborated, and so on (= morphogenesis) continues at their pre-ablation pace; these cycles thus become desynchronized (Costlow 1963a, 1968; Gore 1985). As a result moulting occurs before the morphology of the next normal stage is prepared and an intermediate-stage larva results. Chemicals reported to increase incidence of intermediate stages likely cause similar disruptions to the moult and/or developmental cycles, resulting in these falling out of sync and creating abnormal larvae.

The mechanism by which JH and its analogues induce intermediate stages is less clear because no JH-equivalent neurohormone has been confirmed in crustaceans (Webster 2015). However, the fact that insect JH induces such developmental abnormalities in crustaceans (Table 1) suggests that at the very least some form of JH-like receptor has been retained in the Crustacea, though its function is unclear. Several other neurohormones are known to be involved in moult-cycle control, including MF, CHH, and others (Webster 2015), so it is at least conceivable that some of these, particularly the JH-precursor MF, might also be involved in control of morphogenesis and/or metamorphosis and act similarly to JH. Alternatively, JH might be similar enough to some of the hormones involved in control of moulting and development that it binds to their receptors, and thus acts as an endocrine disruptor that interferes in these cycles similarly to other chemicals discussed. This remains to be confirmed, however (see section 2.2).

Larvae developing through intermediate stages identical to those induced by artificial means have also been reported among normal, intact larvae reared in the laboratory and sampled from the plankton (Table 1). This is important because it shows that these can occur without apparent disruption of physiological cycles controlling moulting and development, and thus these can presumably occur in nature; if so these do not represent developmental anomalies but are instead normal components of larval development in these species.

3.4. Synthesis: Could Intermediates Stages Represent Unknown Plasticity?

The interruptions to moult cycles that have been observed to produce intermediate-stage larvae could easily be analogous to the mechanisms by which larvae with highly-plastic development adjust their developmental pathways (Nelson 1991; Oliphant et al. 2013). Both of these processes give rise to larvae that moult sooner but complete the larval phase and metamorphosis later than is typical and have morphology intermediate between stages occurring in the predominant sequence. Extra larval stages in plastic taxa have been observed to occur naturally among intact, normal larval due to unknown causes or unfavourable naturalistic environments, but the same sorts of extra stages have also been induced to form in some of the same taxa through experimental manipulation or exposure to unnatural toxicants (Table 1). Similarly, the intermediate stages induced to form by eyestalk ablation, exposure to toxicants, etc. have also been observed among normal, intact larvae in the lab and field (Table 1). Therefore, it seems highly likely that induced intermediates represent real cases of developmental plasticity among larvae previously believed to have fixed, canalized development, rather than representing unnatural abnormalities.

Many of the species in which development has been assumed to be fixed are studied mainly in laboratory settings, where conditions are less variable than those encountered by larvae in nature (Anger 2001). As such, perhaps extra stages were only rarely, if ever, observed in prior studies on such species because lab conditions did not extend into those that lead to delayed development in nature. Anger (2001) stated that variable development may be far more prevalent than has thus far been assumed in nature, and the present review of extra and intermediate larval stages certainly suggests that this may be the case. The frequency with which these may occur in nature and causes of occurrence should thus be better quantified to determine if this is true, because if it is studies on species with intermediate stages need to take these factors into consideration to avoid important errors (see sections 3.2.4 and 4.3 for more on this).

4. Development Time of Lobster (*Homarus* spp.) Larvae: A Case Study of Intermediate Stages and Their Potential Significance

4.1. Overview of Homarid Lobster Species

4.1.1. Species Background Information

Extra larval stages of intermediate morphology have been reported repeatedly in homarid lobsters (see Table 1, Figure 1, 2). There are two known extant species in the genus *Homarus* (Astacidea: Nephropidae): the American lobster *Homarus americanus* and the European lobster *H. gammarus* (Wahle et al. 2013). Both of these species, especially the American one, support extremely valuable fisheries throughout their range (Lawton and Lavalli 1995; Wahle et al. 2013). Management of the most productive lobster fisheries attempts to maintain supplies of future fisheries recruits by conserving egg, and thus larval, production (Wahle et al. 2013). As such, much research effort is devoted to the biology and ecology of lobster larvae because any factors impacting their survival, retention, or health can have large, cascading effects on lobster populations and the fisheries they support (Ennis 1995). It will be shown that occurrence of extra larval stages might impact such aspects of larval biology and fisheries biology, and thus should be included in estimates of larval survival, production, and distribution for fisheries management. These species thus form a case study of the potential significance of better understanding of intermediate larval stages to an economically-important crustacean group.

Both *Homarus* species are large benthic marine crustaceans with distinctive large, asymmetrical chelae (claws) and geographic ranges in the North Atlantic Ocean (*H. americanus*: Cape Hatteras, NC, USA to Labrador, Canada; *H. gammarus*: Morocco to Norway, as well as the Black and Mediterranean Seas) spanning several degrees of latitude and a variety of different thermal environments (Wahle et al. 2013). Adults of the two species are very similar in appearance and can only be distinguished by the presence (*H. americanus*) or absence (*H. gammarus*) of ventral spines on the rostrum (Wahle et al. 2013). The larvae are also very similar, although *H. gammarus* larvae are larger on average that those of *H. americanus* (Carlberg et al. 1978); *H. americanus* appears to grow faster, however, and reaches larger sizes as adults overall than *H. gammarus* (Carlberg et al. 1978; Wahle et al. 2013). Larval development is also slightly different among the two species, in that *H.*

gammarus is more developmentally advanced in its last zoeal stage (III) than *H. americanus*, resulting in shorter duration of this stage and less-dramatic advancement in morphology following the moult to the decapodid (i.e., less metamorphic development, Rötzer and Haug 2015). *H. americanus* has been introduced into European waters, where it may hybridize with *H. gammarus* (Agnalt et al. 2015; Carlberg et al. 1978). These species are capable of hybridizing in the lab and producing viable hybrid larvae that combine the larger size of *H. gammarus* with the more rapid growth of *H. americanus*, resulting in larger individuals at the end of development (Carlberg et al. 1978). It is not known whether hybrids are fertile, but if not their occurrence has potential to have strong negative impacts on populations and fisheries of *H. gammarus* (Agnalt et al. 2015).

4.1.2. Usual Larval Development Pathway

Larvae of the genus *Homarus* normally pass through three zoeal stages (hereafter referred to as stages I, II, and III, as in the lobster literature; Charmantier and Aiken 1987; Charmantier et al. 1991; Ennis 1995; Wahle et al. 2013). All three stages have distinctive zoeal features, including dorsal abdominal spines, long exopodites bearing natatory setae on the pereiopods used for locomotion, long posterolateral and medial spines on the telson, short antennal flagella, non-chelate third pereiopods, and relatively passive feeding behaviour and poor horizontal swimming ability (Charmantier and Aiken 1987; Charmantier et al. 1991). Larvae increase in size and undergo dramatic changes in morphology and behaviour as development progresses. Eyestalks and pleopod buds first appear in stage II (they are not present in stage I) and then their development (as well as that of the X-organ-sinus-gland complex) advances in subsequent stages. Uropods appear in stage III (they are absent in stages I and II) and the chelae (first pereiopods) become more well-developed. Phototaxis appears to be strongly positive early in stage I, presumably leading larvae to remain near surface waters in the ocean, but then becomes more negative later in the stage and remains mostly negative or neutral during stages II and III; presumably this leads stage II and III larvae to remain deeper in the water column than in stage I, but still above the bottom because all three larval stages have negative barotaxis and gravitaxis (i.e., move away from excessive depths; Ennis 1995).

At the end of stage III (the "ultimate" or last zoeal stage) a "metamorphic" moult occurs to lobster stage IV (Figure 1, 2; Charmantier et al. 1991), in which zoeal characters are lost and the lobster becomes more similar in behaviour and form to later benthic juvenile and adult stages (V+).

Specifically, dorsal abdominal spines, natatory exopodites of the pereiopods, and medial posterior telson spines are lost, posterolateral telson spines become very short, the third pereiopod becomes chelate, antennal flagella become longer, pleopods are used for locomotion, and feeding behaviour becomes much more active and horizontal swimming ability increases dramatically (Charmantier et al. 1991; Ennis 1995; Lawton and Lavalli 1995). This stage is typically called by the ambiguous term "postlarva" in the lobster literature (Ennis 1995; Lawton and Lavalli 1995; Wahle et al. 2013) but the use of this term in general has been criticized and should be phased out (see section 2.1; Felder et al. 1985; Martin et al. 2014); nonetheless, the proper status of lobster stage IV as a larval or juvenile stage is unclear. On the basis of behaviour early in the stage (swimming with pleopods, nektonic existence, positive phototaxis and negative baro- and gravitaxis) it can be viewed as a last larval stage dramatically different in form from the zoeae, in which case it represents a decapodid (e.g., Ennis 1995; Gore 1985). However, the morphology of stage IV is not really different enough from subsequent juvenile stages to be considered "larval" (see Anger 2001; Felder et al. 1985). Further, later in the stage preceding and after settlement to the benthos phototaxis becomes negative, baro- and gravitaxis become positive, and swimming becomes infrequent (Ennis1995; Lawton and Lavalli 1995). Thus, stage IV could alternatively be considered to represent the first juvenile stage of the lobster lifecycle (Anger 2001; Felder et al. 1985). In the present chapter, a "functional" definition of stage IV as a larval (decapodid) stage will be used because, like the zoeal stages, this stage inhabits the water column, has not yet "recruited" to benthic populations (settled), and has potential to move among different locations; it thus is ecologically closer to a larva than a juvenile for this genus. The moult from stage III to IV can be considered a true morphological metamorphosis (Charmantier et al. 1991), but a second "ecological" or "functional" metamorphosis occurs via the transition of the lobster from a pelagic to benthic existence, which occurs when bottom-seeking and settlement behaviour manifests within stage IV (Cobb et al. 1989; Ennis 1995).

4.1.3. Significance of Larval Development Time to Lobster Biology and Ecology

Moulting rate of lobster larvae is temperature-dependent, such that at higher temperatures the time spent in a given stage (and the duration of the entire larval phase) is generally less than that at lower temperatures (*H. americanus*: MacKenzie 1988; Quinn et al. 2013; further references within

both studies; *H. gammarus*: Schmalenbach and Franke 2010; Figure 3); at very high temperatures, though, moulting may slow again due to thermal stress (Quinn and Rochette 2015). Total zoeal development time from hatch to the moult to stage IV ranges from 10 days (at 22°C) to 56 days (at 10°C) or more in *H. americanus* (MacKenzie 1988; Figure 3) and from 14 days (at 22°C) to 26 days (at 14°C) in *H. gammarus* (Schmalenbach and Franke 2010; Figure 3); stage IV lobsters can also spend considerable time (e.g., 54 days at 10°C; MacKenzie 1988) drifting or swimming if they delay settlement to the end of this stage (Figure 3). The length of time required for larvae to complete development to the settling stage thus controls for how long and how far larvae drift, which has important demographic implications; for example, increasing the length of time spent drifting by 60% could lead to larvae drifting up to 500 km further before settling (Quinn 2014). Larvae that drift for longer time periods are also more likely to die because they have a greater chance of being eaten by predators (Morgan 1990; Vaughn and Allen 2010) or being advected into locations with unfavourable conditions for survival and settlement (Incze and Naimie 2000). Development and drift time of larvae thus strongly impacts the numbers of larvae settling in particular localities, the percentage retention of local egg production and percentage of annual settlement that is locally-produced (i.e., the relative independence versus interdependence of fishing grounds), and population or stock structure of the species (Quinn 2014). Accurate estimates of larval development times are thus crucial for making estimates the aforementioned aspects of lobster stocks, populations, and fisheries.

4.2. Induced Intermediate Stages of Lobster Larvae

4.2.1. Observed Types of Intermediate Stages in Homarid Lobsters and Their Characteristics

In addition to the normal series of three zoeae (I-III) and one functional decapodid (IV) stages, larvae of homarid lobsters have been observed to develop through extra stages with morphology intermediate between other, normally-occurring stages (Table 1, Figure 1, 2). The first report of this phenomenon was by Williamson (1905), who observed a few larvae intermediate in form between stages III and IV among normal *H. gammarus* larvae reared in Scottish hatcheries, which he interpreted as an alternate form of stage III. Templeman (1936) reported that 13 larvae out of 379 otherwise normal *H. americanus* larvae reared to the fourth stage in the laboratory over a three-year period moulted from stage III into "stage IV" larvae with morphology intermediate between stages III and IV. Wells and Sprague (1976) reported occurrence of "intermediate" *H. americanus* larvae, but did not describe these in detail. Tully and Ó Ceidigh (1987) reported an unusual larva of *H. gammarus* found among those collected from nature with plankton tows, which could not be assigned to regular larval stages and appeared intermediate between stages III and IV. A series of studies by G. Charmantier and colleagues (Charmantier and Aiken 1987; Charmantier et al. 1984a, b, 1985a, b, 1988) and others (Snyder and Chang 1986) on *H. americanus* larvae expanded on the nature and occurrence of these intermediate-stage larvae.

Charmantier et al. (1984a) demonstrated aspects of metamorphosis in *H. americanus* by conducting eyestalk ablation on larvae, the effects of which included prevention of physiological changes in hydromineral regulation that normally occur following the moult from stage III to IV (e.g., stage III larvae are isoionic while stage IV switch to being slightly hyperionic, but ablated stage IV larvae remained isoionic to seawater). In subsequent tests of the role of eyestalks and their associated glands (e.g., the X-organ-sinus-gland complex) in osmoregulatory development of lobsters, Charmantier et al. (1984b) noted occurrence of intermediate-stage larvae among ablated larvae when ablation of eyestalks occurred prior to a "critical period": moult stage D_1 (Drach 1949) in stage II (= during stage D_0; see section 2.2). Snyder and Chang (1986) tested the effects of eyestalk ablation on American lobster larval development, and found that moulting rates and the frequency of occurrence of intermediate-stage larvae increased following ablation. Snyder and Chang (1986) also noted that two types of larvae intermediate in form between stages III and IV ("early" = more like stage III and "late" = more like stage IV) occurred. Three "types" of intermediate larval stages occurring following bilateral eyestalk ablation were observed by Charmantier et al. (1985a). These were then described in detail by Charmantier and Aiken (1987). Charmantier et al. (1988) conducted further trials with eyestalk ablation, which demonstrated more rapid moulting in ablated larvae and that the types of intermediate stages produced by ablation depending on the timing of the operation in relation to the aforementioned critical period in stage II (Charmantier et al. 1984b).

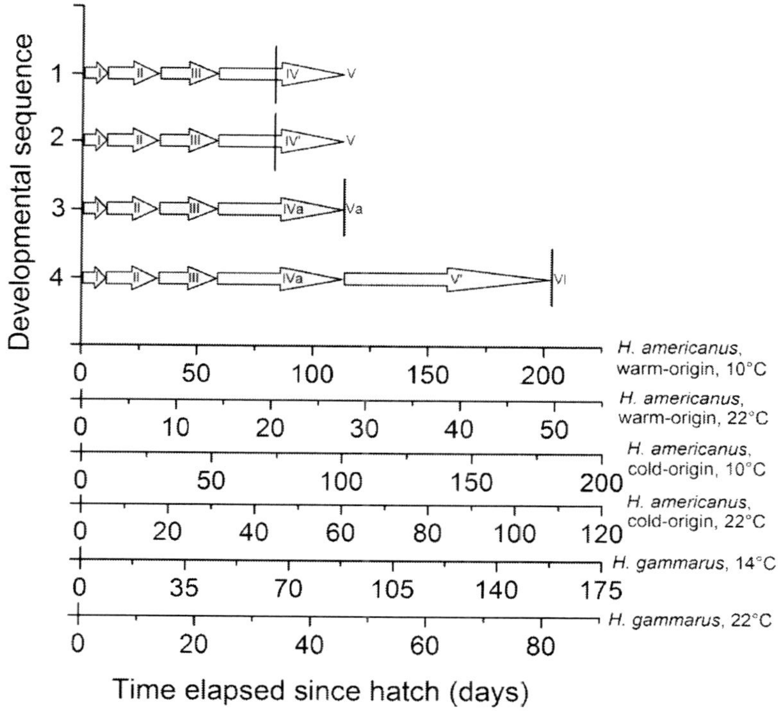

Figure 3. Potential developmental pathways for lobster (*Homarus* spp.) larvae. Each arrow represents a single larval stage (grey = zoea, white = decapodid). Stage numbers not enclosed by arrows are the first true juvenile stage ('Va' = normal stage V with a few larval characters retained; Charmantier and Aiken 1987). Vertical black bars represent the earliest point at which settlement to the benthos can likely occur; this is thought to normally be ~halfway through stage IV (Cobb et al. 1989). Developmental pathways are arranged with the normal and most frequent one at the top (sequence #1), and then alternatives are listed in order of increasing developmental delay. Time required to reach and complete each stage in the normal sequence based on data from MacKenzie (1988), Quinn et al. (2013), and Schmalenbach and Franke (2010). Durations of intermediate stages assumed to be equal to those of equivalent normal stages (IV and V), but these could actually require more or less time (see text, sections 4.3.2 and 4.3.3).

The three intermediate stages that have been observed in larvae of homarid lobsters, particularly the American lobster (Table 1), were designated stages IVa, IV′, and V′ by Charmantier and Aiken (1987) (see Figure 1, 2), and this is the terminology for these stages that has been used since (Charmantier et al. 1988; Ennis 1995). Stages IVa and IV′ occur after the normal stage III and both of these intermediate stages can potentially moult into mostly normal first-stage juveniles (stage V) (Figure 3). However, a few stage IVa larvae moult into the intermediate stage V′, which then moults into a normal second-stage juvenile (stage VI) (Figure 3). These observations thus suggest that four different developmental pathways are available to lobster larvae, two (normal and IV′) consisting of three zoea at different levels of developmental advancement and a decapodid, one with four zoeae and perhaps no decapodid (IVa → V), and one (IVa → V′) with five zoeae and no decapodid (Figure 3). Charmantier et al. (1988) showed that timing of bilateral eyestalk ablation impacted the types of intermediate stage produced, such that when the operation was carried out earlier in stage II the interruption of larval development was stronger. Specifically, ablation long before the D_1 critical point was most likely to result in eventual production of stage IVa larvae that then moulted into stage V′ intermediates, while ablation later more often produced IVa larvae that moulted into stage V; ablation very close to the critical point was more likely to result in production of stage IV′ intermediates (Charmantier et al. 1988).

Stage IVa is the least developmentally advanced of these intermediates and bears several zoeal characters including long telson spines (posteriolateral and medial), dorsal abdominal spines, short antennal flagella, nonchelate third pereiopods, and long exopodites on the pereiopods bearing setae (Figure 1, 2; Charmatier and Aiken 1987); the overall relative size and allometry of this stage is closer to normal stage IV than III, though, and the lengths of exopodites are actually reduced relative to normal stage III (Charmatier and Aiken 1987). Stage IVa also behaves much more like a stage III zoea than a stage IV lobster, most notably in that it lacks bottom-seeking behaviour (Charmantier and Aiken 1987) and thus represents a true, additional larval (zoeal) stage. Stage IV′ is more advanced than stage IVa and largely resembles a stage IV (e.g., same size and shape, long antennal flagellum, similar pereiopods development) except that it still has dorsal abdominal spines and long setae-bearing exopodites on the pereiopods (Figure 1; Charmantier and Aiken 1987); also, though the overall telson shape is similar to normal stage IV and there is no medial spine the posteriolateral spines on the telson are still longer and curved inward relative to those of normal stage IV (Figure 2, 4). Stage IV′ behaves much more like normal stage IV than does stage IVa and even exhibits bottom-seeking behaviour (Charmantier and Aiken 1987; Charmantier et al. 1988); it can thus be considered a less-developed decapodid stage. Charmantier and Aiken (1987) concluded that the intermediate larvae reported and figures by Williamson (1905) and

Templeman (1936) were equivalent to their stages IV′ and IVa, respectively. Intermediate larval stages reported by Snyder and Chang (1986) were also concluded to represent Charmantier and Aiken (1987)'s stages IVa (= "early" intermediates of Snyder and Chang, 1986) and IV′ (= "late"). The larva noted by Tully and Ó Ceidigh (1987) was likely equivalent to Charmantier and Aiken (1987)'s stage IV′ (this chapter).

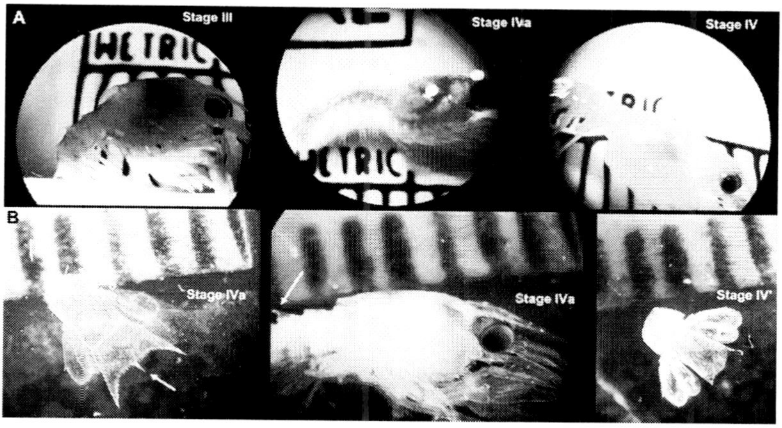

Figure 4. Photographs of normal and intermediate-stage American lobster larvae produced in recent larval rearing studies (unpub. data). (A) Larvae from Quinn et al. (2013), from left to right: normal stage III zoea, intermediate-form stage IVa extra zoea (note dorsal abdominal spines, indicated by white arrows), and normal decapodid (stage IV); photos taken by author. (B) Larvae from Thurlow (2014), from left to right: telson of stage IVa extra zoea (note medial spine on telson and dorsal abdominal spines), cephalothorax and abdomen of same stage IVa larva (note dorsal abdominal spine), and telson of possible intermediate decapodid stage IV′ (note rectangular shape of telson and long, curved posteriolateral spines); photos taken by Samantha Thurlow. Divisions on ruler in background of both (A) and (B) = 1 mm.

Intermediate stage V′ resembles a normal stage V juvenile except that it still bears relatively long spines on the telson, dorsal abdominal spines, and long pereiopodal exopodites with setae (Figure 1, 2; Charmantier and Aiken 1987). Importantly, though, behaviour of stage V′ differs from that of a normal juvenile, or indeed even a normal stage IV decapodid, in that it shows no bottom-seeking behaviour (Charmantier and Aiken 1987) meaning that settlement may not occur until the next stage (VI); this stage may thus be functionally an additional larval (zoeal?) stage. It should also be noted that when a stage IVa larva moults to stage V rather than V′, the resulting lobster is essentially a juvenile in form and behaviour (e.g., it shows bottom-seeking and walking behaviour) except that its telson is intermediate in shape between that of a larva (rectangular) and juvenile (rounded) and it does retain pereiopodal exopodites, but these lack setae (Charmantier and Aiken 1987); these larval features are presumably lost at the moult to stage VI. This could be argued to represent a fourth kind of intermediate stage, (e.g., Va?; Figure 3), but Charmantier and Aiken (1987) considered it to merely be a normal juvenile retaining a few larval characters.

Intermediate larval stages have since been reported and discussed in a few other studies of American lobster (Agnalt et al. 2015; Hall and Bowden 2012; Laufer et al. 2012; Rötzer and Haug 2015), some of which tied their occurrence to environmental stressors (see next section). They were also observed in some recent studies of temperature-dependent larval development rates (Quinn et al. 2013) and geographic differences in morphology of larval swimming appendages (Thurlow 2014) (see Figure 4; unpub. data).

4.2.2. Presumed Causes of Intermediate Stages in Lobsters

By far the most commonly cited cause of intermediate larval stages in homarid lobsters is experimental removal of both eyestalks (bilateral eyestalk ablation: Charmantier and Aiken 1987; Charmantier et al. 1984b, 1985a, 1988; Snyder and Chang 1986; Table 1). As described earlier (section 3.3.3) this procedure removes or damages important components of the neuroendocrine system, reduces production of MIH, and puts the developmental cycle out of sync with the moulting cycle (Costlow 1963a, 1968; Webster 2015). This may also impact processes related to metamorphosis when it occurs in zoea II, wherein many aspects of later development seem to be "decided" based on larval condition and other factors (Anger 2001; Charmantier et al. 1984a, b; Charmantier et al. 1988; Costlow 1963a). Injection of insect JH was also reported to result in production of a few stage IV′ intermediates by Charmantier et al. (1988), suggesting the existence of hormonal pathways and receptors in lobster larvae like those in insects that could be involved in regulating development and metamorphosis (Webster 2015).

Other stressors have also been tied to occurrence of intermediate stages in *Homarus* spp. larvae (Table 1). Templeman (1936) noted that most of the intermediate larvae they observed occurred among larvae reared in less optimal conditions, specifically in seawater of low salinity and temperature. The intermediate-stage larvae observed by Quinn et al. (2013; unpub. data, Figure 4A) and Thurlow (2014; unpub. data, Figure 4B) were observed to occur more frequently in the warmest rearing conditions tested (22°C) and less often in cooler treatments (10, 14.5, and 18°C). As these larvae originated from a relatively cold-water location, it is possible that intermediates were induced by thermal stress effects at temperatures not usually experienced by them in nature. However, larval survival in these studies was highest in general at 22°C so it is more likely that more intermediates occurred at this temperature simply because more larvae lived to moult from stage III. Snyder and Chang (1986) observed that the quality of food provided to larvae during lab rearing mediated whether intermediate-stage larvae occurred, such that even eyestalk-ablated larvae formed intermediates much less often when fed live *Artemia* nauplii (a high-quality food item) than frozen adult *Artemia* (low-quality food). Use of such low-quality food in the studies of Quinn et al. (2013) and Thurlow (2014) may thus have led to occurrence of intermediates. These studies also reared larvae in a semi-automated apparatus that moved culture containers on a regular basis (see Quinn et al. 2013 for details) and thus could have caused injuries to larvae; if eyestalks were injured, then formation of intermediates may have been initiated. Intermediate-stage American lobster larvae were found to occur more frequently among those exposed to human-produced pollutants such as crude oil (Wells and Sprague 1976), toxic potash brine (Charmantier et al. 1985b), and endocrine-disrupting alkylphenols (Laufer et al. 2012). Researchers investigating potential impacts of ocean acidification on lobsters have recently noted that more intermediate-stage larvae occur in low-pH treatments (Hall and Bowden 2012).

The stressors listed above could cause intermediate stages either by directly interfering in moult- and development-cycle regulation in a way analogous to eyestalk ablation (e.g., alkylphenols (Laufer et al. 2012) likely work in this way) or by inducing stress responses in the form of developmental plasticity, as observed extensively among other crustaceans (see Table 1 and section 3.2.3); this topic will be returned to later (section 4.3). Rötzer and Haug (2015) theorized that the more metamorphic development pattern of *H. americanus* is more-derived than the less-metamorphic pattern seen in *H. gammarus*, but under suboptimal conditions this derived pattern (or, more specifically, the hormonal cycles maintaining it?) breaks down and the development pattern reverts to its ancestral form. In studying potential impacts of invasive *H. americanus* on *H. gammarus* fisheries in Norway Agnalt et al. (2015) observed that intermediate larval stages (possibly IVa?) were very frequent among hybrid larvae of the two species, to the point that no clear stage III larvae could be found among hybrid broods (Agnalt et al. 2015). It is possible that in hybrid larvae moult and development cycles are inherently out of sync, resulting in these abnormalities, and if so the long-term viability of hybrids is questionable.

Table 1. Specific examples of occurrence, types, and potential causes of extra and intermediate larval stages among decapod crustaceans (Phylum Arthropoda: Subphylum Crustacea: Class Malacostraca: Subclass Eumalacostraca: Superorder Eucarida: Order Decapoda: Suborder Pleocyemata). Examples were drawn primarily from reviews by Knowlton (1974), Williamson (1982), Gore (1985), and Anger (2001) and from references in other studies cited in this chapter. If the currently-accepted species name or classification differed from that originally reported it has been updated here to agree with modern taxonomic convention

Species	Types and characteristics of extra stages	Cause(s) of occurrence	Reference(s)
Infraorder Anomura: Superfamily Galatheoidea: Family Galatheidae			
Galathea intermedia	Extra zoea V occasionally occurs before decapodid	Occur among normal reared in the lab; causes unknown	Christiansen and Anger 1990
Pleuroncodes planipes	Extra zoea IV with 4-8 "substages" (different degrees of intermediates?)	Occur among normal larvae reared in the lab, and also in nature; varies among broods	Boyd and Johnson 1963
Family Porcellanidae			
Petrolisthes elongatus	"Substages" of zoeae I and II (=intermediates?)	Occur among larvae from nature, causes unknown	Greenwood 1965; Wear 1964a
Petrolisthes novaezelandiae	"Substages" of zoeae I and II (=intermediates?)	Occur among larvae from nature, causes unknown	Greenwood 1965; Wear 1964b
Pisidia longicornis	Rapid moulting through larval stages, juveniles with larval characters (= intermediate stages?)	Bilateral eyestalk ablation	Le Roux 1979
Porcellana spp.	"Substages" of zoeae I and II (=intermediates?)	Occur among larvae from nature, causes unknown	Lebour 1943

Species	**Types and characteristics of extra stages**	**Cause(s) of occurrence**	**Reference(s)**
Superfamily Hippoidea: Family Hippidae			
Emerita talpoida	Extra zoeae between zoeae III and IV, with intermediate morphology	Occur among normal larvae reared in the lab, causes unknown	Johnson and Lewis 1942
Superfamily Paguroidea: Family Coenobitidae			
Coenobita clypeatus	Extra zoeal stages added terminally before decapodid	Occur among normal larvae in the lab; causes unknown	Provenzano 1962b
Family Diogenidae			
Calcinus tibicen	Extra zoeal stages added terminally before decapodid	Occur among normal larvae in the lab; causes unknown	Provenzano 1962a
Infraorder Astacidea: Superfamily Nephropoidea: Family Nephropidae			
Homarus gammarus	Extra intermediate zoea (equivalent to intermediate stage IV′ in *H. americanus*?)	Occur among normal, intact larvae reared in hatcheries; causes unknown Occur among normal, intact larvae in nature from plankton tows	Williamson 1905 Tully and Ó Céidigh 1987
H. americanus x *H. gammarus* hybrids	Extra zoea (equivalent to intermediate stage IVa in *H. americanus*?)	Interspecific hybridization	Agnalt et al. 2015
Homarus americanus	Some individuals moult from zoea III to extra zoea (IVa) instead of to normal decapodid (IV) and may moult into intermediate decapodid or juvenile (V′) or normal juveneile (V) with larval traits; others moult to extra zoea or intermediate decapodid (IV′) that moults to normal juvenile (V)	Bilateral eyestalk ablation; occurrence varies among broods from different females (maternal effects?) Occurrence among normal, intact larvae, possibly in unfavourable conditions (low salinity and/or temperature); occurrence varies among broods from different females Exposure to crude oil constituents Exposure to toxic brine Bilateral eyestalk	Charmantier and Aiken 1987; Charmantier et al. 1984b, 1985a, 1988 Templeman 1936 Wells and Sprague 1976 Charmantier et al. 1985b Snyder and Chang 1986 Charmantier et al. 1988 Laufer et al.

Table 1. (Continued)

Species	Types and characteristics of extra stages	Cause(s) of occurrence	Reference(s)
		ablation; poor food quality Injection of juvenile hormone Alkylphenol-enriched diets Acidic (low pH) rearing conditions	2012 Hall and Bowden 2012
Homarus americanus	As above	Occurrence among normal, intact larvae reared in the lab, causes unknown (Poor food quality, rearing conditions, and/or high temperature?) "Reversion" to ancestral (*H. gammarus*) development pattern under unfavourable conditions	Thurlow 2014 (unpub. data); Quinn et al. 2013 (unpub. data) Rötzer and Haug 2015
Infraorder Brachyura: Superfamily Grapsoidea: Family Plagusiidae			
Euchirograpsus americanus	Less-advanced zoea V moults into extra zoea VI, then decapodid	Occur among normal larvae reared in the lab, causes unknown	Wilson 1980
Family Sesarmidae			
Metasesarma rubripes	Occasionally extra intermediate zoea V occurs before decapodid	Unfavourable conditions during lab rearing	Montú et al. 1990
Sesarma reticulatum	Occasionally extra intermediate zoea V occurs before decapodid	Bilateral eyestalk ablation	Costlow 1966a
Eriocheir sinensis	Extra zoea VI before decapodid; occasionally a second decapodid stage occurs	Low temperature + low salinity	Anger 1991; Montú et al. 1996

Species	Types and characteristics of extra stages	Cause(s) of occurrence	Reference(s)
Neohelice granulata	Two types of zoea IV observed, differing in size and developmental advancement; less-developed type moulted into extra intermediate zoea V, not decapodid	Unfavourable conditions, especially diet (frozen adult *Artemia* vs. live nauplii)	Giménez and Anger 2001; Pestana and Ostensky 1995
Superfamily Homolodromolidea: Family Latreillidae			
Eplumula australiensis	Some larvae lacking usual characters in particular stages (= part of a longer, delayed series of stages?)	Occurrence among normal larvae in the lab, varying among hatchings (maternal effects?)	Williamson 1965
Superfamily Majoidea: Family Epialtidae			
Taliepus dentatus	Extra zoea IV with 4-8 "substages" (different degrees of intermediates?)	Occur among normal larvae reared in the lab, causes unknown	Fageti et al. 1971
Superfamily Portunoidea: Family Portunidae			
Callinectes sapidus	Extra zoea VIII rarely occurs before decapodid	Occur among normal larvae reared in the lab, causes unknown	Costlow and Bookhout 1959
	Zoea IV occasionally moults to a stage intermediate between zoeae IV-V, some of which moult to intermediates between zoeae V-VI and VI-VII; developmental pathways generally variable, with normal zoeal stages sometimes skipped	Occur among normal larvae reared in the lab, causes unknown	Costlow 1965
Macropipus marmoreus	Intermediate stages between zoeae II and	Occur among normal larvae reared in the lab,	Goldstein 1971

Table 1. (Continued)

Species	Types and characteristics of extra stages	Cause(s) of occurrence	Reference(s)
	II, III and IV, and IV and V	causes unknown	
Superfamily Xanthoidea: Family Menippidae			
Menippe mercanaria	Extra zoea VI added before decapodid	Rare occurrence among normal larvae in the lab; occurrence *not* (?) correlated to temperature or salinity Exposure to insecticide (Mirex)	Ong and Costlow 1970 Bookhout and Costlow 1974
Family Panapeidae			
Rhithropanopeus harrisi	Occasionally extra intermediate zoea V occurs before decapodid	Bilateral eyestalk ablation Stressful salinities + exposure to JH analogue methoprene	Costlow 1966b Christiansen et al. 1977
Infraorder Caridea: Superfamily Alpheoidea: Family Alpheidae			
Alpheus heterochaelis	Extra intermediate zoea IV, and sometime V, occur before decapodid or juvenile stage(s); more stages if ablated earlier in zoea II	Bilateral eyestalk ablation	Gross and Knowlton 2002
Family Hippolytidae			
Eualus occultus	Extra zoaea VII-IX (or more) added before decapodid	Occur among normal larvae in nature; causes unknown	Pike and Williamson 1961
Eualus pusiolous	Extra zoea VII added before decapodid	Occur among normal larvae in nature; causes unknown	Pike and Williamson 1961
Exhippolysmata ensirostris	Extra zoeae (20 total, little change	Settlement delay in lab due to lack of suitable	K.N. Sankolli,

Species	Types and characteristics of extra stages	Cause(s) of occurrence	Reference(s)
	after 8th moult, possibly extra instars rather than stages = mark-time moulting?)	substrate?	pers. comm. cited in Williamson 1982, p. 81
Hippolyte inermis	Variable numbers of zoeae between zoea IV (1-4 = zoeae V-VIII) and final zoeal stage (can be zoea VI-IX)	Occur among normal larvae reared in the lab, causes unknown	Le Roux 1963
Lysmata vittata	Extra zoeae (17 total, little change after 8th moult, possibly extra instars rather than stages = mark-time moulting?)	Settlement delay in lab due to lack of suitable substrate?	K.N. Sankolli, pers. comm. cited in Williamson 1982, p. 81
Superfamily Crangonoidea: Family Cragnonidae			
Crangon allmani	Extra zoeal and decapodid stages, with less morphological development between each stage; 4-5 stages (zoeae + decapodid) in larvae from Western Baltic Sea, 5-6 in those from North Sea; some larvae went through up to 8 stages total	Poor food quality, low salinity, and/or extreme temperatures; differences in developmental pattern based on geographic origin of larvae	Criales and Anger 1986
Crangon crangon	Highly variable development: different forms of zoea II occur (from most to least advanced: IIa, IIb, and IIc); four forms of zoea VI; two decapodids, one of which (the least advanced, was IIc)	Inter-individual differences (non-genetic?) within the same brood; induced by poor food quality, low salinity, and/or extreme temperatures	Linck 1995

Table 1. (Continued)

Species	Types and characteristics of extra stages	Cause(s) of occurrence	Reference(s)
	moults either into juvenile or an extra decapodid; most-advanced zoea VI (was IIa?) can skip decapodid stage(s) and moult directly to juvenile		
Superfamily Palaemonoidea: Family Palaemonidae			
Palaemon varians	More larval stages occurred at lower temperatures: 4 stages vs. 5 (2 zoeae + 3 decapodids?) vs. 6-7 in the coldest conditions	Low temperatures	Oliphant et al. 2013
Palaemon debilis	Two "types" of decapodid, one more advanced than the other (= part of a longer, delayed series of stages?)	Occur among normal larvae reared in the lab, causes unknown	Shokita (1977)
Palaemon macrodactylus	1-3 extra intermediate larval stages (zoeae VI-VIII) occurred before decapodid	Bilateral eyestalk ablation	Little 1969
Palaemonetes pugio	Up to 2 extra stages occurred (total = 9-11, normally 7-9)	Exposure to sublethal mercury concentrations	Shealy and Sandifer 1975
Palaemonetes vulgaris	More larval stages occur at higher temperatures	High temperatures	Knowlton 1974
Superfamily Pandaloidea: Family Pandalidae			
Pandalopsis pacificus	Number of stages averages 9 but is variable, can be 18-19 moults	Occur among normal larvae in the lab; causes unknown	Rothlisberg 1980

Species	Types and characteristics of extra stages	Cause(s) of occurrence	Reference(s)
Pandalus spp.	Number of stages varies from 2-13;generally two modes: "fast" (4-7 zoeae) or "slow" (9-13 zoeae)	Occur among normal larvae in the lab; causes unknown	Rothlisberg 1980
Infraorder Thalassinidea: Superfamily Callianassoidea: Family Laomediidae			
Naushonia crangonoides	Abnormal zoea VII less advanced than normal (= part of a longer, delayed series of stages?); one of these moulted three more times (zoeae VIII-X?) but never reached decapodid (= mark-time moulting?)	Occur among normal larvae in the lab and in natural plankton samples; causes unknown	Goy and Provenzano 1978
Family Upogebiidae			
Upogebia deltaura	Extra zoea IV before decapodid, may then moult into an intermediate-form decapodid	Occur among lab-reared larvae, collected from nature in plankton tows; causes unknown	Webb 1919
Upogebia kempi	Number of decapodid stages can be 2 (stages more advanced) or 3 (stages less advanced)	Occur among larvae sampled from nature, causes unknown	Shenoy 1967
Upogebia stellata	Extra zoea IV before decapodid, may then moult into an intermediate-form decapodid	Occur among lab-reared larvae, collected from nature in plankton tows; causes unknown	Webb 1919

Additional factors that have been tied to occurrence of intermediate larval stages in lobsters are genetics, maternal effects and perhaps rearing densities. Templeman (1936) noted that larvae of particular hatchings (i.e., produced by the same female) tended to produce more intermediates than those of others. Charmantier and Aiken (1987) observed that intermediate stages could not be induced to form in larvae from particular females' broods, even following ablation. Thus, certain larvae may be more or less predisposed to go through intermediate stages than others. It is unclear whether this is due to genetic differences among larvae and/or a result of differential maternal conditioning of larvae during embryonic development (e.g., larvae that develop from eggs that receive less yolk from a less-healthy or small female may be more likely to produce intermediates than those more well-provisioned by larger, healthier females). Intermediate stages observed by Quinn et al. (2013) were more prevalent among larvae reared individually (1 larva per 35 cm^3 cell) than among those reared in group culture (starting densities = 300 stage I larvae per 60 L kreisels; unpub. data), which may suggest presence or absence of conspecifics influences the likelihood of intermediates to occur. This may be related to effects of interactions among individuals on moult and growth cycles, as has been observed for juvenile lobsters (Lawton and Lavalli 1995).

Importantly, intermediate stages have also been observed without any clear cause among non-treated (i.e., no eyestalk ablation) lobster larvae (Table 1); these were noted by Williamson (1905) and Charmantier and Aiken (1987). The potential causes for intermediates suggested by Templeman (1936), or for Quinn et al. (2013) and Thurlow (2014) (this chapter), are also largely conjectural because these studies were not designed to test for causes of intermediates. Tully and Ó Céidigh (1987) also observed a possible intermediate-stage larva in nature, suggesting that these can and do occur outside of the laboratory.

4.3. Implications of Intermediates If They Represent a Normal Aspect of Development

4.3.1. Importance of Potential Developmental Plasticity Among Lobsters

Intermediate stage IVa occurs quite frequently (9-52%) among treated lobster larvae (Table 2), as does the subsequent stage V′ when observed (~24%; Table 2), while Stage IV′ is comparatively rare in most cases (3-13%, Table 2). Reported incidences of intermediate-stages of unknown cause among untreated or natural lobster larvae have, conversely, been quite low thus far (mostly <1%, but up to ~13%, see Table 2). Based on this it could be concluded that these intermediates are of minimal importance in nature, so therefore estimating larval development, dispersal, survival, recruitment, and so on for fisheries research on lobsters on the basis of the "normal" developmental pattern (stages I-IV) provides a good approximation to reality. Indeed this is the perspective taken by most studies, which view intermediates more as experimental curiosities elucidating aspects of moulting and development rather than as real components of the lobster life cycle (e.g., Charmantier and Aiken 1987; Ennis 1995). However, this is in no way certain.

As discussed earlier (section 3.2 and Table 1) many crustacean species – possibly the majority of them (Oliphant et al. 2013) – show some degree of developmental plasticity and variable numbers and types of larval stages, which may be an essential adaptation to persistence of species with potentially long larval lives in variable environments (Gore 1985). In more-derived groups, like the Nephropidae (containing *Homarus*) and Brachyura, development observed in the majority of cases seems to be less plastic and more canalized (i.e., fixed at a particular mode or sequence of stages; Anger 2001; Gore 1985). However, the occurrence of more plastic development among even these groups has been observed more and more over time, leading to changes in the way carcinologists view crustacean developmental patterns (Anger 2001; Oliphant et al. 2013). Ultimately, it seems that extra intermediate-form larval stages induced artificially in the lab also occur in nature and represent a normal component of development in many different crustaceans (see section 3.4), likely including homarid lobsters.

It should be noted that the vast majority of larval samples in nature are taken from areas within the middle of the geographic or thermal ranges of species, where they presumably experience more "optimal" conditions resulting in the species' "normal" sequence of larval stages (Oliphant et al. 2013). For instance, American lobster larvae examined in previous work came from areas well-within the centre of the species' range (MacKenzie 1988; Quinn et al. 2013), while study of larvae from more extreme areas near the edges of the range (e.g., North Carolina (very warm, extreme south) and Labrador (very cold, extreme north)) has been rare (Ennis 1995). In species with more variable developmental sequences alternative pathways (in the form of extra or skipped and different stages) occur much more often near the limits of the species' range, where suboptimal conditions are more prevalent than in the more "optimal" middle of their range (e.g., Criales and Anger 1986; Oliphant et al. 2013). Thus, it is possible that occurrence of intermediate-stage lobster larvae is more common in more "extreme" portions of the species' range than the middle regions where most sampling effort has been concentrated (i.e., locations of larger fisheries). This possibility should be investigated. Additionally, knowledge of intermediate stages can be relatively limited among lobster researchers (pers. obs.), so it is possible that intermediates have occurred during numerous previous studies in the lab and nature but not been observed and/or reported. To the naked eye, the overall form and behaviour of a stage IVa or IV′ larva is very similar to that of a normal stage IV and clearly different from those of stage III larvae (pers. obs.); unless these larvae are then subjected to close observation under magnification (which is not always done) they will not be clearly distinguishable from a normal stage IV and will be reported as such (e.g., Quinn et al. 2013; Thurlow 2014). It is thus quite possible that the occurrence of intermediate stages among lobster larvae in general, and particularly in nature, has been vastly underestimated.

4.3.2. Observed and Estimated Development Time Through Intermediate Stages

If intermediate stages are a normal component of lobster larval development in at least part (if not all) of the species' range, and if they actually occur among a large proportion of larvae there, then they need to be accounted for when estimating larval ecology of the species. Calculating connectivity among populations and densities of settlers (= potential fisheries recruits) is usually done by simulating drift of larvae in oceanographic fields (= models). These initiate drift at the beginning of stage I and then terminate it at some point during stage IV (usually beginning halfway through the stage, peaking ~2/3 through the stage, and ending at the moult to stage V; Cobb et al. 1989; Quinn 2014). The development times of the normal sequence of stages I-IV at different temperatures are fairly well-known (MacKenzie 1988; Quinn et al. 2013; Schmalenbach and Franke 2010), so these can be used to control drift times of larvae in such simulations. The basis of larval dispersal modeling depends strongly on accurate simulations of larval development times at different temperatures (e.g., Quinn 2014); specifically, longer development and drift results in different densities of settlers per area, different stock structure, and potentially lower overall larval survival (Quinn 2014). However, it is uncertain whether the same relationships can be applied to larvae that go through intermediate stages. This is a particularly important issue with regard to intermediate stages IVa and V′, because neither of these stages shows settlement behaviour (Charmantier and Aiken 1987) so their occurrence thus extends the larval phase by two full stages (i.e., settlement occurs at or after the moult to stage VI rather than by the end of stage IV; Figure 3). It is also uncertain how long completion of these two stages takes at different temperatures, so their duration cannot even be modeled at this point. Stage IV′ is less problematic because this stage will settle, and so does not increase the number of larval stages, but it is still possible that its duration differs somehow from that of normal stage IV.

Table 2. Frequency of occurrence of intermediate-stage larvae of lobsters (*Homarus* spp.) in previous studies from which data could be extracted. If multiple experiments were carried out in a particular study (e.g., Charmantier et al. 1988) totals in this table were calculated from all larvae surviving the moult from stage III or IVa across all experiments. Treated = larvae operated on (eyestalks ablated) or intentionally exposed to some factor (toxicants, etc.) (see Table 1 for details); Untreated = occurrence among normal, intact larvae due to unknown cases

Study	Treated	Untreated
Charmantier and Aiken 1987; Charmantier et al. 1988[1]	IVa: 461/1032 (52%) IV′: 28/1032 (3%) V′: 31/145 (21.4%)	IVa: 10-15/5000 (0.2-0.3%) IV′: 5-10/5000 (0.1-0.2%) V′: 0/5000? (0%?)
Charmantier et al. 1985b	IVa + IV′: ~10%	
Charmantier et al. 1988[2]	IV′: 6/45 (13.3%)	-
Laufer et al. 2012	22/35 (62%)	0/177 (0%)
Quinn et al. 2013 (unpub. data)	-	IVa: Indiv[3] = 3/72 (4.2%), Group[4] = 1/262 (0.4%) IV′: Indiv[3] = 7/72 (9.7%), Group[4] = 5/262 (1.9%)
Snyder and Chang 1986	IVa: 74/161 (46%) IV′: 15/161 (9.3%)	IVa + IV′: 0/92 (0%)
Templeman 1936	-	13/379 (3.4%)
Thurlow 2014 (unpub. data)	-	IVa: 1/133 (0.8%) IV′: 7/133 (5.3%)
Tully and Ó Céidigh 1987	-	1/50 (2%)
Wells and Sprague 1976	10/521 (9.2%)	1/52 (1.9%)

[1]Treatment = Bilateral eyestalk ablation; [2]Treatment = Injection with insect juvenile hormone (JH); [3]Individualy-reared larvae; [4]Larvae reared in group culture.

A few of the studies that induced intermediates using bilateral eyestalk ablation reported development times of larvae (Snyder and Chang 1986; Charmantier et al. 1988). Ablation during stage II was reported by Snyder and Chang (1986) to result in a slightly shorter (mean duration ± SD = 4.1 ± 0.1 versus 4.3 ± 0.2 days at 18-20°C) duration of this stage versus intact controls, as well as even greater abbreviation of stage III (5.4 ± 0.7 (n = 28) versus 6.2 ± 0.5 (31) or 5.9 ± 0.6 (22) days at 18-20°C) and "subsequent stages" (results not reported). Charmantier et al. (1988) observed that bilateral eyestalk ablation in stages I-III generally decreased development time through these and subsequent stages. Total times taken to reach the fourth (IV, IVa, or IV′) and fifth (V or V′) larval stages ranged from 12-17 and 42-56 days, respectively, in control, intact larvae at 20°C, while ablated larvae took 10-17 and 23-30 days to reach these stages (Charmantier et al. 1988). Development times through stages I-III of larvae injected with JH by Charmantier et al. (1988) did not differ from those of control larvae, however. Unfortunately, Charmantier et al. (1988) did not report development times through stages IVa and IV′ separately from those of normal stage IV larvae, so whether development rates of these intermediates differed from normal stage IV cannot be ascertained.

Table 3. Development times through larval stages I-III at different temperatures of individual American lobster larvae reared by Quinn et al. (2013) that moulted to intermediate larval stages (IVa or IV′) (unpub. data)

Temperature (°C)	Rearing conditions	Type of intermediate stage produced	Development time (days)			
			Stage I	Stage II	Stage III	Total time to fourth stage
10.3	Individual	IVa	5	14	18	37
14.5	Individual	IVa	8	8	17	33
17.5	Individual	IVa	5	10	8	23
22.2	Individual	IV′	5	7	13	25
22.2	Individual	IV′	6	6	7	19
22.2	Individual	IV′	3	7	6	16
22.2	Individual	IV′	3	7	7	17
22.2	Individual	IV′	3	5	9	17
22.2	Individual	IV′	3	6	8	17
22.2	Individual	IV′	4	5	7	16
14.5	Group	IV′				29
17.5	Group	IVa				20
17.5	Group	IV′				21
17.5	Group	IV′				21
22.2	Group	IV′				16
22.2	Group	IV′				14

Note that group-rearing of larvae was not carried out at 10.3°C and observation of individual stage durations was not possible in group-rearing. These data are plotted against development times of larvae that moulted to normal stage IV in Figures 5 and 6.

Larvae moulting into intermediate stages as a result of exposure to toxicants, in contrast, have been reported to show slower moulting than control larvae. Larvae exposed to crude oil, several of which moulted to intermediate stages, required more days (total time = 13.3-18.6 days at 17-20°C) to develop through stages I-III than control ones (14.2 days at 17-20°C) (Wells and Sprague 1976). A similar slowing of development through stages I-III relative to controls was also reported in larvae fed alkylphenol-enriched diets (21-24 days vs. 19 days at 16-18°C), which also produced more intermediates (Laufer et al. 2012). Durations of resulting intermediate stages were not reported in these studies.

New previously unpublished data are presented in this chapter from the larval rearing study of Quinn et al. (2013) (Figure 4-6, Table 3). Larvae reared individually and in groups at different constant temperatures (10.3, 14.5, 17.5, and 22.2°) in that study that completed the moult from stage III to the next stage were examined in detail under a dissecting microscope. Several intermediate stages were identified in this way (Figure 4-6, Table 3) and their development times through preceding larval stages were then compared with those of larvae that moulted into normal stage IV larvae at the same temperatures. Development times could be compared through each of stages I, II, and III for individually-reared larvae (Figure 5) and for total duration of stages I-III for individually- and group-reared larvae (Figure 6). Development times of larvae that moulted into intermediates through stages I and III did not differ from those that moulted to normal stage IV at any temperature (independent samples *t*-tests for each stage and temperature, all $p > 0.05$; Figure 5). There were significant differences in development time through stage II, but these did not follow any consistent pattern across all temperatures; specifically, intermediates had spent less time than normal larvae developing through stage II at 10 and 18°C, but more time at 14.5 and 22°C (*t*-tests, $p < 0.05$; Figure 5). Total development times of intermediates through stages I-III were either slightly longer than, or similar to, those of normal stage IVs for all temperatures and both types of rearing, such that overall differences were non-significant (*t*-tests, $p > 0.05$; Figure 6). When stage IVa and IV′ co-occurred in the same temperature (17.5°C) and rearing conditions (group) there was a slight tendency for stage I-III development of stage IVa larva to be ~1 day longer than that of IV′ larvae (Figure 6), but there was not sufficient replication to test this trend statistically.

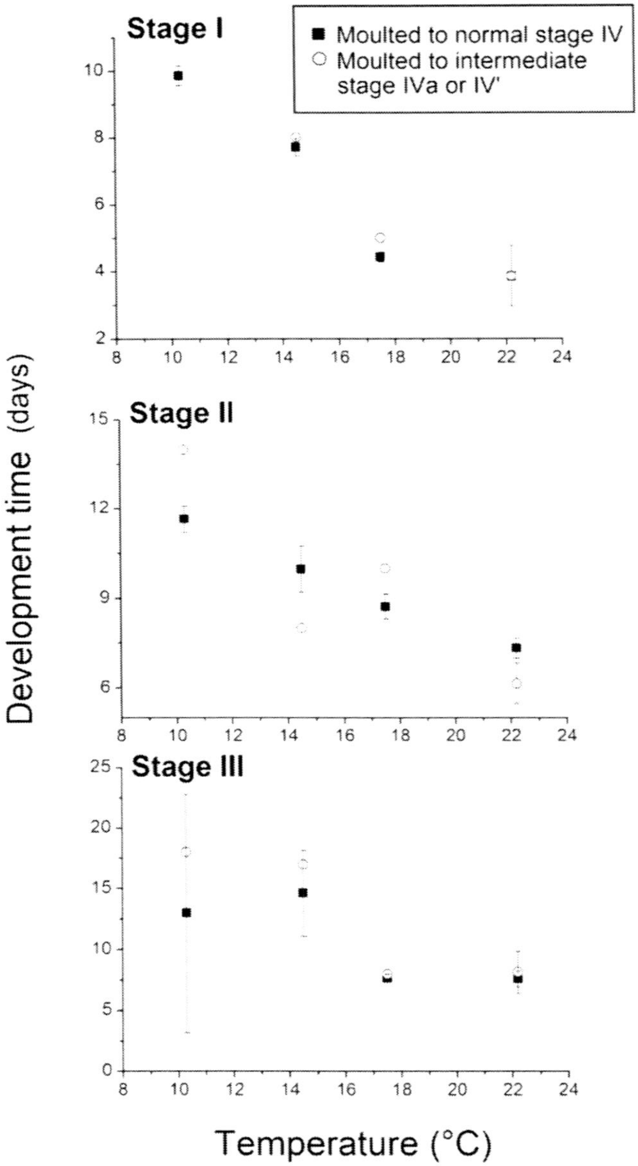

Figure 5. Comparison of development time through each of larval stages I, II, and III between American lobster larvae that moulted to normal stage IV (black squares) versus those that moulted to intermediate stages (IVa and IV′, white circles) at different temperatures in the study by Quinn et al. (2013, unpub. data). Data points are means ± 95% confidence intervals.

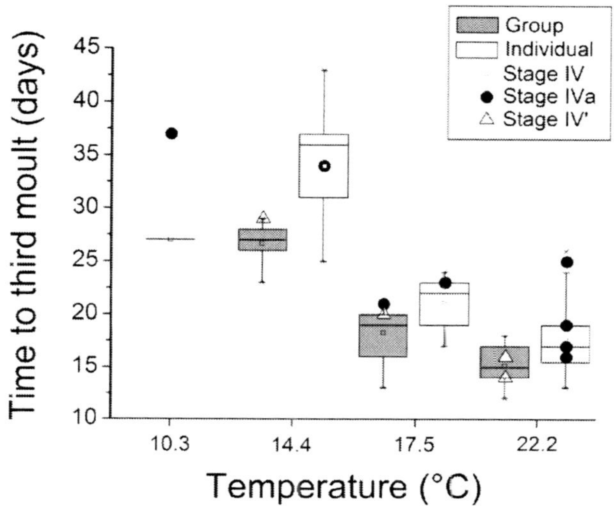

Figure 6. Total development times from hatch to third moult (= to fourth stage) of American lobster larvae reared by Quinn et al. (2013 and unpub. data). Black circles represent individual larvae that moulted from stage III to intermediate stage IVa and white triangles are those that moulted to intermediate stage IV′. Box-and-whisker plots are the distributions of development times of all larvae that reached normal stage IV in group (gray boxes) and individual (white boxes) rearing at different temperatures (boxes = 25th, 50th (median), and 75^{tho}%tiles; whiskers = 5th and 95^{tho}%tiles; 'x' = 1st and 99^{tho}%tiles; square = mean). Note that group rearing was not carried out at 10.3°C and only 2 larvae (1 stage IV, 1 stage IVa) reached the fourth stage in individual-rearing at this temperature.

It can also be noted from the data in Figure 6 that development times of group-reared larvae were generally and on-average significantly (independent samples *t*-tests for each temperature, all $p < 0.05$) less than those of individually-reared larvae at the same temperatures, implying suboptimal conditions in individual rearing. However, the range of group-reared development times did fall within that of individually-reared ones, specifically ≤ median (50th%tile) times for individuals (Figure 6). Cannibalism is frequently observed in group-rearing of lobsters, in which larvae that moult sooner than others attack and eat their slower-developing siblings when they moult (Sastry and Zeitlin-Hale 1977); this results in potential longer development times being removed from the dataset for group-reared larvae. This occurred in the study of Quinn et al. (2013) and is likely the reason for differences between rearings reported here (Figure 6; unpub. data, pers. obs.)

Templeman (1936) reported development time for one of the intermediate-stage (IVa) larvae produced unintentionally in their study. This larva moulted to the next stage (V?) after 6.5 days at 19-20°C, while comparable normal stage IVs required 12 days on average (n = 8) to do so; thus, this intermediate larva spent less time in this stage than did normal stage IV larvae. Importantly, Templeman (1936) is the only published report of actual development time of a lobster larva through an intermediate stage.

4.3.3. Hypothesised Development Times of Intermediates and Their Implications

Based on the above, it is difficult to make any conclusions regarding the durations of intermediate larval stages IVa, IV′, and V′. Ablation of eyestalks is known to accelerate moulting rates in arthropods in general (e.g., Charmantier et al. 1985a; Costlow 1966a; Le Roux 1979) but not always (see Costlow 1966b), and toxicants and JH usually slow moulting, even without producing intermediates (e.g., Laufer et al. 2012; Webster 2015). Therefore, development times reported for intermediates produced by these means, if faster or slower than normal (and even this is uncertain because duration of intermediates is almost never reported) may not be applicable to those of intermediates that occur naturally. Slower or similar development times of larvae moulting into intermediates through preceding stages than those of normal larvae could imply that the duration of IVa and IV′ would follow the same trend (i.e., duration ≥ that of normal stage IV), but this is not guaranteed. The shorter duration of stage IVa than IV reported by Templeman (1936) is intriguing, but as this was only a single larva (n = 1) the general applicability of this result is highly uncertain. There thus exist a number of possibilities, most of which could greatly impact development and drift of larvae as simulated in dispersal and life history models. These are listed and described below:

(1) *Duration of intermediate stages > regular stage IV:* This could have the greatest potential impact. If all intermediate stages take longer to be completed than normal stage IV, then settlement to the benthos of larvae developing through non-settling stages IVa and then V′ before finally settling at juvenile stage VI (Figure 3) would be markedly delayed. Settlement of larvae developing through stages IVa or IV′ → V (Figure 3) would also be slightly delayed relative to normal larvae.

(2) *Duration of intermediate stages = regular stage IV:* This outcome would still have a large impact on one type of intermediate

developmental sequence, but not others. Settlement of larvae developing through stages IVa → V′ → VI would be delayed (e.g., by up to 117 days, Figure 3) days due to the insertion of extra non-settling larval stages, but not by as much as in the previous case (#1). Settlement of larvae developing through stages IVa → V would also be slightly delayed (e.g., by up to 27 days, Figure 3), assuming that IVa intermediates do not settle (Charmantier and Aiken 1987; Figure 3). Settlement of larvae developing through stages IV′ → V would not be delayed in this case (Figure 3).

(3) *Duration of intermediate stages < regular stage IV:* The impacts of this possibility would depend on the degree to which intermediate stages are abbreviated relative to normal stage IV. If abbreviation is slight, then settlement of larvae that develop through stages IVa → V′ → VI would still be delayed relative to normal larvae, but not as much as in cases #1 and 2; this would match patterns observed by Oliphant et al. (2013) for extra stages in the shrimp *Palaemon varians*. However, it is conceivable (if perhaps unlikely) that abbreviation of intermediate stage duration could be sufficient in magnitude to actually cause the overall time to settlement to be equal to or even slightly shorter than that in normal stage IV larvae, even with the addition of extra pre-settlement stages. Regardless of the degree of abbreviation, larvae developing through stages IVa or IV′ → V would in this case settle sooner than normal larvae.

The above list of possible outcomes is of course not comprehensive, and there are other possible outcomes. For instance, there could be differences in development rate among types of intermediates which would mediate these outcomes. This is quite likely, given that more delayed and intermediate developmental sequences (IVa → V′ → VI) tend to result from earlier disruption of moult, developmental, and perhaps metamorphic cycles by eyestalk ablation and cause greater acceleration of moulting than less delayed sequences (IVa → V, IV′ → V) (Charmantier et al. 1988; Figure 3).

There are two main potential outcomes in the above list: either larval life is extended or shortened. An argument could be made in favour of either case being adaptive, or not. Delayed completion of the larval phase is potentially advantageous if it allows a larva to ultimately become a larger, better-conditioned juvenile than possible if it settled at the "normal" point under suboptimal conditions (Oliphant et al. 2013), but this is disadvantageous because it increases the chance of planktonic mortality (e.g., by predation) during the larval phase (Vaughn and Allen 2010). Abbreviated larval life could be advantageous because it reduces the probability of mortality in the plankton, but is disadvantageous if settlement must occur at a small size, under suboptimal conditions (Taylor 2003). In both cases there is a trade-off between post-settlement (juvenile) survival (favoured by the former) and pre-settlement (larval) survival (favoured by the latter). It is not clear which of these selective forces acts more strongly on *Homarus* larvae (Ennis 1995; Lawton and Lavalli 1995; Wahle et al. 2013). Better understanding of potential developmental plasticity in these species and the nature of extra intermediate larval stages could help to clarify this question, in addition to providing valuable information for the study of early life history of lobsters in general. It is important that such information be obtained in new studies. In the next section, potential experiments that could be carried out to test the above predictions are described in the hopes that these might stimulate future study in this area.

4.4. Outline of Future Research Needs

Further research is thus needed before the potential impact of extra intermediate stages on homarid lobster life history, populations, and fisheries can be assessed. A crucial knowledge gap concerns the time spent by larvae in intermediate stages. This information should be readily obtainable through a few simple, directed experiments.

As a first step, bilateral eyestalk ablation could be performed to produce intermediates in larvae reared individually at different temperatures (e.g., those comparable to previous studies, such as 10-22°C; MacKenzie 1988). Sibling larvae not treated with ablation (controls) could be reared at the same temperatures. All larvae could then be examined closely after moulting to their fourth and fifth stage to determine whether they are in normal stage IV or V or intermediate stages. Then, durations of intermediate stages could be observed and compared to those of normal larvae and juveniles. Presumably, if these stages differ in length from that of normal stage IV then this will be detectable even among ablated larvae.

A follow-up study should then attempt to observe durations of many intermediate stages at different temperatures without ablation or other treatment to induce intermediates and then compare these to normal stage IV durations. This would be logistically challenging, however, given the low frequencies (<1-13%; Table 2) with which intermediate stages occur among intact larvae, so such a study would need to use very high numbers of larvae (e.g., Charmantier and Aiken (1987) observed intermediates among 0.3-0.5% of intact larvae (Table 2), so one could expect to produce 30-50 intermediates out of 10,000 larvae at these frequencies). However, many lobster hatcheries are likely equipped to rear many thousands of larvae at a time, so such a study is possible in principle. It may also be useful to attempt rearing intact larvae at more extreme temperatures (<10°C, >22°C) than have been tested previously (e.g., MacKenzie 1988; Quinn et al. 2013; Schmalenbach and Franke 2010) because in other decapods more extreme temperatures result in different developmental sequences and stage numbers (Table 1; Oliphant et al. 2013); this may also apply to lobsters, and if so might produce far higher numbers of intact intermediates for testing and analyses than have been produced in previous studies.

It will also be important to better quantify the frequency with which intermediate-stage lobster larvae occur in nature to determine their ecological significance (or lack thereof); this could be accomplished through plankton tows and surveys, combined with careful examination of presumptive stage IV lobsters obtained for intermediate morphology. If these stages almost never occur in nature, then further study of them may be superfluous; but, if they do occur regularly and among large enough proportions of natural larvae – and the above review of occurrence of such extra stages in other crustaceans (section 3) suggests that they likely do – then they need to be given more consideration in future studies of lobster larvae.

CONCLUSION

The prevailing view among carcinologists has been (and in some fields still is) that each crustacean species passes through a fixed series of larval stages during its development, but in actual fact many species pass through different numbers and types of stages depending on conditions (food level, environmental temperature, etc.) experienced during development. Indeed, developmental plasticity appears to be the rule, rather than the exception, among the Crustacea. Extra stages with intermediate characteristics have been induced in many taxa traditionally thought to have fixed development, for example lobsters (*Homarus* spp.), and these stages may represent real, alternative developmental pathways for such species in nature. If so, the causes and frequency of occurrence of such stages needs to be better quantified. The extent to which occurrence of extra/intermediate stages lengthens development time of larvae, for instance, could have strong impacts on predictions of population connectivity, larval supply, etc. used to inform fisheries science and management. Many uncertainties remain, however, regarding the nature and characters of these stages, which should be dealt with in future research.

In conclusion, the following questions have been identified during the present review, and should be pursued in future studies of the larval biology of crustaceans in general, and homarid lobsters specifically: What neuroendocrine hormones and cycles are involved in control of development and morphogenesis and how do these interact with the moult cycle and its constituents? How are larval and post-larval survival impacted by different developmental pathways, are there cases where selection for one overrules that for the other, and if so is this related to the degree of developmental plasticity within a species? With what frequency do extra larval stages and/or alternative developmental pathways occur in nature? What factors influence the occurrence of extra stages, and how? How are development times through each larval stage, particularly extra intermediate ones, and through the larval phase as a whole impacted by occurrence of extra stages? Answering these questions will greatly improve the ability of carcinologists to make accurate estimates of larval ecology and population demography in many species, including those supporting major fisheries, as well as to predict how these will change following future perturbations.

ACKNOWLEDGMENTS

I wish to thank Erin Miller for reviewing an early draft abstract for this chapter; Patrick Ouellet, Rémy Rochette, and Bernard Sainte-Marie for permission to present unpublished data from our larval development study (Quinn et al. 2013); Marthe Larsen Haarr and Samantha Thurlow for providing pictures and information on fourth-stage larvae from their study (Thurlow 2014); Jeff Clements for assistance obtaining obscure but vital papers for this review; the University of New Brunswick (UNB), Saint John Campus and Library for providing access to most of the literature reviewed in this study; and Keiji and Kaneyo Nozawa for hosting me during the preparation of the manuscript. I am also grateful to Nadya Gotsiridze-Columbus and Nova Science Publishers, Inc., for inviting me to submit this chapter for inclusion in the present collection. While working on this chapter I was supported by funding from a National Sciences and Engineering Research Council Canada Graduate Scholarship Doctoral (NSERC CGS D), New Brunswick Innovation

Foundation Graduate Scholarship, UNB Board of Governor's Merit Award for Graduate Studies, and Vaughan Graduate Fellowship in Marine Sciences.

Figures 1 and 2 were reproduced from the paper published in the *Journal of Crustacean Biology*, 1987, 7(3): 525-535, by authors G. Charmantier and D.E. Aiken, © The Crustacean Society, 1987, published by Brill NV, Leiden, with permission from the publisher.

References

Agnalt, A. L., Grefsrud, E., Farestveit, E. & Jørstad, K. (2015). Implications of export of live American lobster (*Homarus americanus*) to Norway: crossbreeding and ecological implications. In: *Proceedings of The American Lobster in a Changing Ecosystem II: A US-Canada Science Symposium, 3–6 November 2015, Charlottetown, Prince Edward Island.* Prince Edward Island Fishermen's Association, Charlottetown, PE, p. 25 (abstract).

Anger, K. (1991). Effects of temperature and salinity on the larval development of the Chinese mitten crab *Eriocheir sinensis* (Decapoda: Grapsidae). *Mar Ecol Prog Ser*, *72*, 103-110.

Anger, K. (2001). *Crustacean Issues 12: The Biology of Decapod Crustacean Larvae*. Rotterdam, The Netherlands: A. A. Balkema, 417 pp.

Anger, K. (2006). Contributions of larval biology to crustacean research: a review. *Invertebr Reprod Dev*, *49*, 175-205.

Bookhout, C. G. & Costlow, J. D. (1974). Larval development of *Portunus spinicarpus* reared in the laboratory. *Bull Mar Sci*, *24*, 20-51.

Boyd, C. M. & Johnson, M. W. (1963). Variations in the larval stages of a decapod crustacean, *Pleuronectes planipes* (Galatheidae). *Biol Bull*, *124*, 141-152.

Broad, A. C. (1957). The relationship between diet and larval development of *Palaemonetes*. *Biol Bull*, *112*, 162-170.

Brooks, W. K. & Herrick, F. H. (1892). The embryology and metamorphosis of the Macrura. *Mem Nat Acad Sci*, *5*, 321-576.

Carlberg, J. M., Van Olst, J. C. & Ford, R. F. (1978). A comparison of larval and juvenile stages of the lobsters *Homarus americanus*, *Homarus gammarus*, and their hybrid. *Proc Ann Meet – World Maricult Soc*, *9*, 109-122.

Carli, A. (1978). Chiave analitica per l'identificazione degli stadi larvali planctonici e del primo stadio postlarvale di *Palemon elegans*, *P. serratus*,

P. xiphias (Crustacea Decapoda). *Mem Biol Mar Oceanogr n.s.*, (8)*5*, 115-122.

Charmantier, G. & Aiken, D. E. (1987). Intermediate larval and postlarval stages of *Homarus americanus* H. Milne Edwards, 1837 (Crustacea: Decapoda). *J Crustac Biol*, *7*, 525-535.

Charmantier, G., Charmantier-Daures, M. & Aiken, D. E. (1984a). Neuroendocrine control of hydromineral regulation in the American lobster *Homarus americanus* H. Milne Edwards, 1837 (Crustacea, Decapoda). 2. Larval and post-larval stages. *Gen Comp Endocr*, *54*, 20-34.

Charmantier, G., Charmantier-Daures, M. & Aiken, D. E. (1984b). Variation des capacités osmorégulatrices des larves et post-larves de *Homarus americanus* H. Milne-Edwards, 1837 (Crustacea, Decapoda). *CR Hebd Acad Sci, Paris*, *299*, 863-866.

Charmantier, G., Charmantier-Daures, M. & Aiken, D. E. (1985a). Intervention des pédoncules oculaires dans le contrôle de la metamorphose chez *Homarus americanus* H. Milne-Edwards, 1837 (Crustacea, Decapoda). *CR Hebd Acad Sci, Paris*, *300*, 271-276.

Charmantier, G., Charmantier-Daures, M. & Aiken, D. E. (1988). Larval development and metamorphosis of the American lobster *Homarus americanus* (Crustacea, Decapoda): effect of eyestalk ablation and juvenile hormone injection. *Gen Comp Endocr*, *70*, 319-333.

Charmantier, G., Charmantier-Daures, M. & Aiken, D. E. (1991). Metamorphosis in the lobster *Homarus* (Decapoda): a review. *J Crustac Biol*, *11*, 481–495.

Charmantier, G., Charmantier-Daures, M. & Young-Lai, W. W. (1985b). Lethal and sublethal effects of potash brine on different stages of the lobster, *Homarus americanus*. *Can Tech Rep Fish Aquat Sci*, *1344*, 1-13.

Christiansen, M. E. & Anger, K. (1990). Complete larval development of *Galathea intermedia* Lilljeborg reared in laboratory culture (Anomura: Galatheidae). *J Crustac Biol*, *10*, 87-111.

Christiansen, M. E., Costolow, J. D. & Monroe, R. J. (1977). Effects of the juvenile hormone mimic ZR-515 (Altosid) on larval development of the mud crab *Rhithropanopeus harrisii* in various salinities and cyclic temperatures. *Mar Biol*, *39*, 269-279.

Cobb, J. S., Wang, D. & Campbell, D. B. (1989). Timing of settlement by postlarval lobsters (*Homarus americanus*): field and labaroatory evidence. *J Crustac Biol*, *9*, 60-66.

Costlow, J. D. (1962). The effect of eye-stalk extirpation on metamorphosis of megalops of the crab *Callinectes sapidus* Rathburn. *Am Zool*, *2*, 401-402.

Costlow, J. D. (1963a). Regeneration and metamorphosis in larvae of the blue crab, *Callinectes sapidus* Rathburn. *J Exp Zool*, *152*, 219-227.

Costlow, J. D. (1963b). The effect of eyestalk extirpation on metamorphosis of megalops of the blue crab, *Callinectes sapidus* Rathburn. *Gen Comp Endocr*, *3*, 120-130.

Costlow, J. D. (1965). Variability in larval stages of the blue crab *Callinectes sapidus*. *Biol Bull*, *128*, 58-66.

Costlow, J. D. (1966a). The effect of eyestalk extirpation on larval development of the crab, *Sesarma reticulatum* Say. In: *Some Contemporary Studies in Marine Science*. Barnes H (ed.). London: George Allen and Urwin, pp. 209-224.

Costlow, J. D. (1966b). The effect of eyestalk extirpation on larval development of the mud crab, *Rhithropanopeus harrisii* (Gould). *Gen Comp Endocr*, *7*, 255-274.

Costlow, J. D. (1968). Metamorphosis in crustaceans. In: *Metamorphosis: A Problem in Developmental Biology*. Etkin W, Gilbert LI (eds.). New York: Appleton-Century-Crofts, pp. 3-41.

Costlow, J. D. & Bookhout, C. G. (1959). The larval development of *Callinectes sapidus* reared in the laboratory. *Biol Bull*, *116*, 373-396.

Criales, M. M. & Anger, K. (1986). Experimental studies on the larval development of the shrimps *Crangon Crangon* and *C. allmanni*. *Helgoländer Meeresun*, *40*, 241-265.

Dakin, W. J. (1938). The habits and life-history of a penaieid prawn (*Penaeus plebejus* Hesse). *Proc Zool Soc London* (A), 1938, 163-183.

Drach, P. (1949). Périodes de mue chez une crustacé à croissance lente *Xantho floridus* Montagu dans le district septentrional de son aire de repartition. *CR Acad Sci Paris*, *228*, 128-130.

Ennis, G. P. (1995). Larval and postlarval ecology. In: *Biology of the Lobster* Homarus americanus. Factor JR (ed.). New York: Academic Press Inc., pp. 23-46.

Esperk, T., Tammaru, T. & Nylin, S. (2007). Intraspecific variability in number of larval instars in insects. *J Econ Entomol*, *100*, 627-645.

Fageti, G., Fageti, E. & Campodonico, I. (1971). Desarrollo larval en el laboratorio de *Taliepus dentatus* (Milne Edwards) (Crustacea Brachyura: Majidae Acanthonychinae). *Revta Biol Mar, Valparaiso*, *14*, 1-14.

Faxon, W. (1879). On some young stages in the development of *Hippa*, *Porcellana* and *Pinnixa*. *Bull Mus Comp Zool*, *5*, 253-268.

Felder, D. L., Martin, J. W. & Goy, J. W. (1985). Patterns in early postlarval development of decapods. In: *Crustacean Issues 2: Larval Growth*.

Wenner AM (ed.). Rotterdam, The Netherlands: A. A. Balkema, pp. 163-225.

Fraser, F. C. (1936). On the development and distribution of the young stages of krill (*Euphausia superba*). *'Discovery' Rept*, *14*, 1-192.

Freeman, J. A. & Costlow, J. D. (1980). The molt cycle and its hormonal control in *Rhithropanopeus harrisii* larvae. *Dev Biol*, *74*, 479-485.

Giménez, L. L. & Anger, K. (2001). Relationships among egg size, embryonic development, and larval biomass in the estuarine crab *Chasmagnathus granulata* Dana, 1851. *J Exp Mar Biol Ecol*, *260*, 241-257.

Goldstein, B. (1971). Développement larvaire de *Macropipus marmoreus* (Leach) en laboratoire (Crustacea, Decapoda, Portunidae). *Bull Mus Natn Hist, Nat, Paris*, (2)*42*, 919-943.

Gore, R. H. (1985). Moulting and growth in decapod larvae. In: *Crustacean Issues 2: Larval Growth*. Wenner AM (ed.). Rotterdam, The Netherlands: A.A. Balkema, pp. 1-65.

Goy, J. W. & Provenzano, A. J. (1978). Larval development of the rare burrowing mud shrimp *Naushonia crangonoides* Kingsley (Decapoda: Thalassinidea: Laomeiidae). *Biol Bull*, *154*, 241-261.

Greenwood, J. G. (1965). The larval development of *Petrolisthes elongatus* (H. Milne Edwards) and *Petrolisthes novaezelandiae* Filhol (Anomura, Porcellanidae) with notes on breeding. *Crustaceana*, *8*, 285-307.

Gross, P. S. & Knowlton, R. E. (2002). Morphological variation among larval-postlarval intermediates produced by eyestalk ablation in the snapping shrimp *Alpheus heterochaelis* Say. *Biol Bull*, *202*, 43-52.

Gurney, R. (1936). Larvae of decapod Crustacea. Part I: Stenopidea. Part II: Amphionidae. Part III: Phyllosoma. *'Discovery' Rep*, *12*, 377-440.

Gurney, R. (1942). *The Larvae of Decapod Crustacea*. London: Ray Society, 306 pp.

Gurney, R. & Lebour, M. V. (1941). On the larvae of certain Crustacea Macrura, mainly from Bermuda. *J Linn Soc London, Zool*, *41*, 89-181.

Hall, J. J. & Bowden, T. J. (2012). Impact on larval developmental of chronic exposure to a reduced pH environment in the American lobster (*Homarus americanus*). In: *Proceedings of The American Lobster in a Changing Ecosystem: A US-Canada Science Symposium, 27–30 November 2012, Portland, Maine*. Maine Sea Grant, University of Maine, Walpole, ME, p. 16 (abstract).

Hartnoll, R. G. (1982). Growth. In: *The Biology of the Crustacea*, Vol. *2: Embryology, Morphology, and Genetics*. Abele LG (ed.). New York: Academic Press, pp. 111-196.

Haug, C., Haug, J. T. & Waloszek, D. (2009). Morphology and ontogeny of the Upper Jurassic mantis shrimp *Spinosculda ehrlichi* n. gen. sp. from southern Germany. *Palaeodiversity*, *2*, 9111-118.

Haug, J. T., Haug, C., Waloszek, D., Maas, A., Wulf, M. & Schweigert, G. (2009). Development in Mesozoic scyllarids and implications for the evolution of Achelata (Reptantia, Decapoda, Crustacea). *Palaeodiversity*, 2, 97-110.

Haug, J. T., Haug, C., Waloszek, D. & Schweigert, G. (2011). The importance of lithographic limestones for revealing ontogenies in fossil crustaceans. *Swiss J Geosci*, *104* (Supp. 1), S85-S98.

Higgins, L. E. & Rankin, M. A. (1996). Different pathways in arthropod postembryonic development. *Evolution*, *50*, 573-582.

Incze, L. & Naimie, C. E. (2000). Modelling the transport of lobster (*Homarus americanus*) larvae and postlarvae in the Gulf of Maine. *Fish Oceanogr*, *9*, 99-113.

Johnson, M. W. & Lewis, W. M. (1942). Pelagic larval stages of the sand crabs *Emerita analoga* (Stimpson), *Blepharipoda occidentalis* Randall, and *Lepidopa myops* Stimpson. *Biol Bull*, *83*, 67-87.

Knowlton, R. E. (1970). Abbreviated larval development in *Alpheus heterochaelis* Say (Crustacea, Decapoda, Caridea) from coastal North Carolina. *Am Zool*, *10*, 542-543.

Knowlton, R. E. (1973). Larval development of the snapping shrimp *Alpheus heterochaelis* Say reared in the laboratory. *J Nat Hist*, *7*, 273-306.

Knowlton, R. E. (1974). Larval developmental processes and controlling factors in decapod Crustacea, with emphasis on Caridea. *Thalassia Jugoslav*, *10*, 138-158.

Laufer, H., Baclaski, B. & Koehn, U. (2012). Alkylphenols affect lobster (*Homarus americanus*) larval survival, molting and metamorphosis. *Invertebr Reprod Dev*, *56*, 66-71.

Lawton, P. & Lavalli, K. L. (1995). Postlarval, juvenile, adolescent, and adult ecology. In: *Biology of the Lobster* Homarus americanus. Factor JR (ed.). New York: Academic Press Inc., pp. 47-88.

Le Roux, A. (1963). Contribution á l'étude du développement larvaire d'*Hippolyte inermis* Leach (Crustacé Décapode Macroure). *CR Séanc Acad Sci, Paris*, *256*, 3499-3501.

Le Roux, A. (1966). Contribution á l'étude du développement larvaire de *Clibanarius erythropus* (Latreille) (Crustacé Décapode Anomoure Diogénidé). *Cah Biol Mar*, *7*, 225-230.

Le Roux, A. (1979). Effects de l'ablation des pédoncules oculaires sur le developpement de *Pisidia longicornis* (Crustacé, Décapode, Anomoure). *Ann Endocrinol*, *40*, 79-80.

Lebour, M. V. (1928). The larval stages of the Plymouth Brachyura. *Proc Zool Soc Lond*, 1928, 473-560.

Lebour, M. V. (1930). The larval stages of *Caridion*, with a description of a new species, *C. steveni. Proc Zool Soc Lond*, 1930, 181-193.

Lebour, M. V. (1943). The larvae of the genus *Porcellana* (Crustacea Decapoda) and related forms. *J Mar Biol Assoc UK*, *25*, 721-737.

Linck, B. M. (1995). Einfluß von Temperatur und Salzgehalt auf die Larven der Nordseegarnele *Crangon Crangon.* M.Sc. Thesis, Universität Oldenburg, Germany, 174 pp.

Little, G. (1969). The larval development of the shrimp, *Palaemon macrodactylus* Rathburn reared in the laboratory, and the effect of eyestalk extirpation on development. *Crustaceana*, *17*, 69-87.

Makarov, R. (1968). On the larval development of the genus *Scelrocrangon* G.O. Sars (Caidea, Crangonidae). *Crustaceana* Suppl., *2*, 27-37.

Makarov, R. R. & Maslennikov, V. V. (1981). Ecology of larval development of the crustacean *Euphausia superba*. Change in dominant larval forms as a function of environmental conditions. *Mar Ecol Prog Ser*, *4*, 265-271.

Martin, J. W., Olesen, J. & Høeg, J. T. (2014). Introduction. In: *Atlas of Crustacean Larvae*. Martin JW, Olesen J, Høeg JT (eds.). Baltimore, MD, USA: Johns Hopkins University Press, pp. 1-7.

Montú, M., Anger, K. & Bakker, C. d. (1990). Variability in the larval development of *Metasesarma rubripes* (Decapoda, Grapsidae) reared in the laboratory. *Nerítica, Pontal Do Sul (Brazil)*, *5*, 113-118.

Montú, M., Anger, K. & Bakker, S. d. (1996). Larval development of the Chinese mitten crab *Eriocheir sinensis* H. Milne-Edwards (Decapoda: Grapsidae) reared in the laboratory. *Helgoländer Meeresun*, *50*, 223-252.

Morgan, S. G. (1990). Impact of planktivorous fishes on dispersal, hatching, and morphology of estuarine crab larvae. *Ecology*, *71*, 1639-1652.

Nelson, K. (1991). Scheduling of reproduction in relation to molting and growth in malacostracan crustaceans. In: *Crustacean Issues 7: Crustacean Egg Production*. Wenner AM, Kuris A (eds.). Rotterdam, The Netherlands: A. A. Balkema, pp. 77-113.

Ngoc-Ho, N. (1976). The larval development of the prawns *Macrobranchium equidens* and *Macrobranchium* sp. (Decapoda: Palaemonidae), reared in the laboratory. *J Zool*, *181*, 439-464.

Nichols, P. R. & Keeney, P. M. (1963). Crab larvae (*Callinectes*) in plankton collections from cruises of M/V Theodore N. Gill, South Atlantic Coast of the U.S. 1953-1954. *US Fish Wld S, Spec Sci Rep Fish*, *448*, 14 pp.

Oliphant, A., Hauton, C. & Thatje, S. (2013). The implications of temperature-mediated plasticity in larval instar number for development within a marine invertebrate, the shrimp *Palaemonetes varians*. *PLoS ONE*, *8*(9), e75785.

Ong, K. S. & Costlow, J. D. (1970). The effect of salinity and temperature on the larval development of the stone crab, *Menippe mercenaria* (Say), reared in the laboratory. *Chesapeake Sci*, *11*, 16-29.

Passano, L. M. (1961). The regulation of crustacean metamorphosis. *Am Zool*, *1*, 89-95.

Pestana, D. & Ostrensky, A. (1995). Occurrence of an alternative pathway in the larval development of the crab *Chasmagnathus granulata* Dana, 1851 under laboratory conditions. *Hydrobiologia*, *306*, 33-40.

Phillips, B. F. & Sastry, A. N. (1980). Larval ecology. In: *The Biology and Management of Lobsters* Volume *II: Ecology and Management*. Cobb JS, Phillips BF (eds.). New York: Academic Press, pp. 11-57.

Pike, R. B. & Williamson, D. I. (1961). The larvae of *Sirontocaris* and related genera (Decapoda, Hippolytidae). *Crustaceana*, *2*, 187-208.

Provenzano, A. J. (1962a). The larval development of *Calcinus tibicen* (Herbst) (Crustacea, Anomura) in the laboratory. *Biol Bull*, *123*, 179-202.

Provenzano, A. J. (1962b). The larval development of the tropical land hermit *Coenobita clypeatus* (Herbst) in the laboratory. *Crustaceana*, *4*, 207-228.

Quinn, B. K. (2014). Assessing potential influence of larval development time and drift on large-scale spatial connectivity of American lobster (*Homarus americanus*). MSc thesis, University of New Brunswick, Saint John, NB, Canada.

Quinn, B. K. & Rochette, R. (2015). Potential effect of variation of water temperature on development time of American lobster larvae. *ICES J Mar Sci*, *72* (Suppl 1), i79-i90.

Quinn, B. K., Rochette, R., Ouellet, P. & Sainte-Marie, B. (2013). Effect of temperature on development rate of larvae from cold-water American lobster (*Homarus americanus*). *J Crustac Biol*, *33*, 527-536.

Rochanaburanan, T. & Williamson, D. I. (1976). Laboratory survival of larvae of *Palaemon elegans* Rathke and other caridean shrimps in relation to their distribution and ecology. *Estuar Coast Mar Sci*, *4*, 83-91.

Rothlisberg, P. C. (1980). A complete larval description of *Pandalus jordani* Rathburn (Decapoda, Pandalidae) and its relation to other members of the genus *Pandalus*. *Crustaceana*, *38*, 19-48.

Rötzer, M. A. I. N. & Haug, J. T. (2015). Larval development of the European lobster and how small heterochronic shifts lead to a more pronounced metamorphosis. *Int J Zool*, 2015, 345172, 17 pp.

Safranek, L. & Williams, C. M. (1984a). Critical weights for metamorphosis in the tobacco hornworm, *Manduca sexta*. *Biol Bull*, *167*, 555-567.

Safranek, L. & Williams, C. M. (1984b). Determinants of larval molt induction in the tobacco hornworm, *Manduca sexta*. *Biol Bull*, *167*, 568-578.

Sandifer, P. A. (1973a). Larvae of the burrowing shrimp, *Upogebia affinis* (Crustacea, Decapoda, Upogebiidae), from Virginia plankton. *Chesapeake Sci*, *14*, 98-104.

Sandifer, P. A. (1973b). Mud shrimp (*Callianassa*) larvae (Crustacea, Decapoda, Callianassidae) from the Virginia plankton. *Chesapeake Sci*, *14*, 149-159.

Sandifer, P. A. & Smith, T. I. J. (1979). Possible significance of variation in the larval development of palaemonid shrimp. *J Exp Mar Biol Ecol*, *39*, 55-64.

Sastry, A. N. & Zeitlin-Hale, L. (1977). Survival of communally reared larval and juvenile lobsters, *Homarus americanus*. *Mar Biol*, *39*, 297-303.

Schmalenbach, I. & Franke, H. D. (2010). Potential impact of climate warming on the recruitment of an economically and ecologically important species, the European lobster (*Homarus gammarus*) at Helgoland, North Sea. *Mar Biol*, *157*, 1127-1135.

Shealy, M. H. & Sandifer, P. A. (1975). Effects of mercury on survival and development of the larval grass shrimp *Palaemonetes vulgaris*. *Mar Biol*, *33*, 7-16.

Shenoy, S. (1967). Studies on larval development in Anomura (Crustacea, Decapoda). II. *Proc Symp Crustacean Mar Biol Assoc India*, *Part II*, 777-804.

Shokita, S. (1977). Larval development of palaemonid prawn, *Palaemon (Palaemon) debilis* Dana from the Ryuku Islands. *Bull Sci Engng Div Univ Ryukyus*, *23*, 57-76.

Snyder, M. J. & Chang, E. S. (1986). Effects of eyestalk ablation on larval molting rates and morphological development of the American lobster, *Homarus americanus*. *Biol Bull*, *170*, 232-243.

Strathmann, R. R. (1985). Feeding and nonfeeding larval development and life-history evolution in marine invertebrates. *Ann Rev Ecol Syst*, *16*, 339-361.

Taylor, D. L. (2003). Size-dependent predation on post-settlement winter flounder *Pseudopleuronectes americanus* by sand shrimp *Crangon septemspinosa*. *Mar Ecol Prog Ser*, *263*, 197-215.

Templeman, W. (1936). Fourth stage larvae of *Homarus americanus* intermediate in form between normal third and fourth stages. *J Biol Board Can*, *2*, 349-354.

Thurlow, S. (2014). Effect of female geographic origin and size on the size and swimming structures of American lobster (*Homarus Americanus*) larvae. BSc thesis, University of New Brunswick, Saint John, NB, Canada.

Tighe-Ford, D. J. (1977). Effects of juvenile hormone analogues on larval metamorphosis in the barnacle *Elminius modestus* Darwin (Crustacea: Cirripedia). *J Exp Mar Biol Ecol*, *26*, 163-176.

Tully, O. & Ó Céidigh, P. (1987). The seasonal and diel distribution of lobster larvae (*Homarus gammarus* (Linnaeus)) in the neuston of Galway Bay. *J Cons int Explor Mer*, *44*, 5-9.

Vaughn, D. & Allen, J. D. (2010). The peril of the plankton. *Integr Comp Biol*, *50*, 552-570.

Wahle, R. A., Castro, K. M., Tully, O. & Cobb, J. S. (2013). *Homarus*. In: *Lobsters: Biology, Management, Aquaculture and Fisheries* (2nd Ed.). Phillips BF (ed.). Oxford, UK: John Wiley and Sons Ltd., pp. 221-258.

Wear, (1964a). Larvae of *Petrolisthes elongatus* (Milne Edwards, 1837) (Crustacea, Decpoda, Anomura). *Trans R Soc NZ, Zool*, *5*, 39-53.

Wear, R. G. (1964b). Larvae of *Petrolisthes novaezelandiae* Filhol, 1885 (Crustacea, Decpoda, Anomura). *Trans R Soc NZ, Zool*, *4*, 229-244.

Webb, G. E. (1919). The development of the species of *Upogenbia* from Plymouth Sound. *J Mar Biol Assoc UK*, *12*, 81-134.

Webster, S. G. (2015). Endocrinology of molting. In: *The Natural History of the Crustacea*, Vol. *4*, *Physiology*. Chang ES, Thiel M (eds.). New York: Oxford University Press, p. 1-35.

Wells, P. G. & Sprague, J. B. (1976). Effects of crude oil on American lobster (*Homarus americanus*) larvae in the laboratory. *J Fish Res Board Can*, *33*, 1604-1614.

Williamson, D. I. (1965). Some larval stages of three Australian crabs belonging to the families Homolidae and Raninidae, and observations on the affinities of these families (Crustacea: Decapoda). *Aust J Mar Fresh Res*, *16*, 369-398.

Williamson, D. I. (1976). Larvae of Stenopodidae (Crustacea, Decapoda) from the Indian Ocean. *J Nat Hist*, *10*, 497-509.

Williamson, D. I. (1982). Larval morphology and diversity. In: *The Biology of the Crustacea*, Vol. *2*, *Embryology*, *Morphology*, *and Genetics*. Abele LG (ed.). New York: Academic Press, pp. 43-110.

Williamson, H. C. (1905). A contribution to the life history of the lobster (*Homarus vulgaris*). *Ann Rep Fish Board Scot 1904*, *Part 3*, *23*, 65-107.

Wilson, K. A. (1980). Studies on decapod Crustacea from the Indian River region of Florida. XV. The larval development under laboratory conditions of *Euchirograpsus americanus* A. Milne Edwards 1880 (Crustacea, Decapoda: Grapsidae) with notes on grapsid subfamilial larval characters. *Bull Mar Sci*, *30*, 756-775.

Yatsuzuka, K. (1962). Studies on the artificial rearing of the larval Brachyura, especially of the larval Blue Crab, *Neptunus pelagicus* Linnaeus. *Rep Usa Mar Biol Stn Kochi Univ*, *9*, 1-4.

In: Crustaceans
Editor: Victor Alvarado

ISBN: 978-1-63484-890-9

Chapter 3

A COMPARATIVE STUDY OF LIPOSOLUBLE VITAMIN EFFECTS IN DECAPOD CRUSTACEAN WITH EMPHASIS IN ARGENTINE SHRIMP SPECIES

Nair de los Angeles Pereira[1], María Victoria Sarasa[2] and Analia Fernández-Gimenez[1,*]

[1]Instituto de Investigaciones Marinas y Costeras,
Facultad de Ciencias Exactas y Naturales,
Universidad Nacional de Mar del Plata-Consejo Nacional de Investigaciones Científicas y Técnicas, Mar del Plata, Argentina
[2]Museo Municipal de Ciencias Naturales Lorenzo Scaglia,
Municipalidad de General Pueyrredón,
Mar del Plata, Argentina

ABSTRACT

Vitamins are important for normal growth and survival of living organisms and they are distributed in feedstuffs in small quantities. There are two groups of vitamins: water-soluble and fat-soluble. The later includes vitamins A, D, E and K. This chapter summarizes several studies carried out with the Argentine penaeoid shrimp species *Pleoticus muelleri*

* E-mail: fgimenez@mdp.edu.ar.

and *Artemesia longinaris,* and their comparison with other Decapoda species, regarding liposoluble vitamins, A, D, E and K, in the diets and metabolic responses. Argentine shrimp species are distributed along South American coast line, from 29°S to 49°S. Growth, survival, and histological analysis were used as monitors of the nutritional value derived from vitamin deficiency. Liposoluble vitamins are also related to stress and antioxidant defense of shrimp. It compared these results with those reported for other crustaceans. This information increased knowledge about biology of Argentine shrimp species and this would help in the formulation of suitable feeds for shrimp. There is, thus need to formulate an effective diet for shrimp in culture conditions, to provide not only macro nutrients, including protein and lipid, but also micronutrients such as vitamins that support optimal growth of shrimp.

Keywords: nutrition, physiology, shrimp, vitamins

INTRODUCTION

The continued growth of the aquatic farming industry depends on the development of nutritious feeds that allow a sustainable aquaculture; a challenge that can only be met with updated information on the nutritional requirements of shrimp. Research in topics such as, digestive physiology and metabolism, are needed to know how to use nutritional data in feed formulation. Thus, optimized feeds and supplements are suitable to improve growth, health and survival of different ontogenetic stages of shrimp; as well as help to reduce pollution in aquatic environment (NCR 2011, 1).

Farmed shrimp have a specific requirement for at least essential nutrients so; effort should be focused on further improvements in feed formulation techniques and on formulating rations on the basis of individual digestible nutrient levels rather than on crude gross nutrient levels (Tacon et al. 2009, 1).

In addition to proteins, lipids and carbohydrates, minerals and vitamins are also important for normal growth and survival of crustaceans (Liñán-Cabello et al. 2002, 299). Feed formulations for shrimp use vitamin levels based on information derived from studies on related species. Widely divergent quantities of vitamins, generally much higher levels, have been added into shrimp feeds using vitamin premixes (Reddy et al. 1999, 267). On other hand, changes in metabolic requirements can be induced by variations in food composition; for example, increasing the lipid level in diet, increases the demand for vitamins probably as a physiological response (Conklin 1989, 291).

In this chapter, we report the results of several studies carried out with the Argentine penaeoid species, *Pleoticus muelleri* and *Artemesia longinaris* regarding liposoluble vitamins in diets using metabolic, physiological and histological responses as indicator of vitamin deficiency. We also considered liposoluble vitamins in relation with stress and antioxidant defense and these results are compared with those reported for other crustaceans.

Vitamins are complex organic substances of small molecular size. They are distributed in feedstuffs in small quantities and form a distinct entity from other major and minor food components. There are two groups of vitamins: water-soluble and fat-soluble. The later includes vitamins, A, D, E and K (Conklin 1989, 288). Its importance as essential constituents in the diets came to light in the early part of this century; however nutrition research on aquatic organisms was partly due to the inherent problems posed by the aquatic medium (vitamin leaching). As vitamin requirements remain unknown, formulated feeds may still be deficient in certain vitamins even after supplementation (Raj 1987, 2).

Some author developed a vitamin diet test for fish containing crystalline vitamins, casein, dextrin and oils with crab meal or dried liver as the source for the antianemic factor. Subsequently, the vitamin free casein purified diet, which is widely used as the standard diet for determining the qualitative and quantitative requirements of a number of aquatic species, has been developed (Halver 1957, 246). Conklin (1981, 148) proposed a reference diet with vitamin free casein and Castell (1989, 340) formulated another diet containing a crab protein concentrate, as the principal protein source yielded satisfactory growth and survival of marine and freshwater crustaceans.

Dietary vitamin requirements in cultured crustaceans have been originally demonstrated with dietary trials of growth and survival. However, it is also possible that vitamin deficiencies and over fortifications may be manifested through changes in the midgut gland cells (Reddy et al. 1999, 267; Fernández-Gimenez et al. 2004, 1177).

The midgut gland is the largest organ in decapod crustaceans and has many biological functions, including synthesis and secretion of digestive enzymes, absorption of digested products, maintenance of mineral reserves and organic substances, lipid and carbohydrate metabolism, distribution of stored reserves during the molt cycle and catabolism of some organic compounds. It is a conspicuous and bilobulated gland composed of simple, blind ended tubules, and leads to the pyloric stomach. Each tubule has three defined zones: distal, middle and proximal and is surrounded by connective tissue. There can be identified four cell types as well: E cells (embryonic), F (fibrillar), R (resorptive) and B (secretory) form the epithelium (Ceccaldi 1997, 265).

The digestion process is cyclical, and the specific role of the different cellular types has been determined through morphological, ultrastructural, histochemical, and immunohistochemical studies.

It is important to recognize the structure of the midgut gland because of the role it plays in the metabolism and the rapid histological changes it suffers in response to different physiological demands and environmental changes. It has been noted that this organ undergoes histological and histochemical modifications in response to physiological demands, such as molt reproduction and pollution (Al-Mohanna and Nott 1989, 538). Rodriguez Souza et al. (1996, 133) suggest that histological criteria constitute a practical mean for the preliminary assessment of the acceptability and nutritional value of diets for the crustaceans.

Argentine Shrimp Species

There are two commercial penaeoid species in the Argentine waters, *P. muelleri* and *A. longinaris*. They live in waters with temperatures between 6 and 23°C and salinities from 31.5 to 33.5 ppt. The shrimp *P. muelleri* is an open thelycum species, distributed along the South American coastal waters from Río de Janeiro, Brazil (23°S) to Santa Cruz, Argentina (49°45′S). Males reach 50g and females 90g total weight, and their spawning capacity can be over 360,000 eggs per female. This species also has a high commercial value; it is remarkable that over 80% of the catches are exported to Spain and Italy. The shrimp *A. longinaris* is a closed thelycum species that shows a similar distribution in the northern area but reaches only the latitude of 43°S. It is most abundant in the southern part of Buenos Aires (Mar del Plata, Bahia Blanca) and the northern part of Patagonia (Rawson). Males reach 12g and females 15g, this species is traded mainly in the internal market (Fenucci et al. 2009, 164).

Both species have seasonal and annual catches fluctuations, for what it is important to establish the feasibility of culturing them on commercial basis to provide continuous supplies to the market.

Previous studies have been done, mainly with *P. muelleri*, on different aspects on the biology, nutrition, maturation; large-scale larval culture and pond grow out (Díaz et al. 2004, 3189; 2011, 2 Fenucci et al. 2009, 148; Fernández Gimenez et al. 2010, 11).

VITAMIN A AND CAROTENOIDS

Vitamin A is the generic descriptor for compounds with the qualitative biological activity of all-*trans*-retinol, and can be found in animals as retinol, retinal and retinyl esters. Retinoid, or their precursors, is essential in the diet of vertebrates and their function as a component of retinal pigment in the eye has long been known (Bearer-Rogers et al. 2001, 727).

Fisher et al. (1952, 256) found very large quantities of retinoids in the eyes of euphausiid species but only small amounts in the rest of the body. In other crustacean species, retinoid were either absent or constituted 1mg g^{-1} wet mass. In eight species of Penaeidea gender, some pelagic or deep-sea species had no retinoids in the eyes and only trace amounts or none in the body. Shallow-water species such as the northern brown shrimp *Farfantepenaeus aztecus* had none in the body, but 4 mg g^{-1} in the eyes (Fisher et al. 1957, 215), a similar situation was found in the rusty crayfish *Orconectes rusticus* (Wolfe and Cornwell 1965, 207).

Several researches showed vitamin A have a more fundamental function in gene transcription and are essential in cell division, differentiation and reproduction (Combs 1998, 103).

In crustaceans, this vitamin is involved in mucopolysaccharide metabolism, visual pigments and maintenance of epithelial, calcium transport across some membranes, embryonic development and in cellular and subcellular membrane integrity (Tacon 1991, 19). Vitamin A has also been related to the normal ovarian development of kuruma shrimp *Marsupenaeus japonicus* (Alava et al. 1993, 1235) whiteleg shrimp *Litopenaeus vannamei* (Liñán-Cabello et al. 2002, 299), and fleshy prawn *Fenneropenaeus chinensis* (Mengqing et al. 2004, 295).

Many studies have evaluated the dietary essentiality of vitamin A for penaeids (Kanazawa 1984, 127; He et al. 1992, 177; Reddy et al. 1999, 267; Fernández-Gimenez 2002, 134). He et al. (1992, 181) demonstrated that at least 130 mg vitamin A in feed of *L. vannamei* prevent deleterious effects on growth, on other hand Liang and Ji (1998, 88) recommended about 40 to 60 mg vitamin A kg^{-1} to achieve good survival of *F. chinensis'* larvae. Otherwise, it was demonstrated in *P. muelleri* and *A. longinaris*, that there is no relation between the weight gain and survival of shrimp and the level of vitamin A in feeds (Fernández-Gimenez 2002, 135; Fernández-Gimenez et al. 2008, 487).

Several studies have been carried out to characterize different physiological stages of the shrimp *P. muelleri* and *A. longinaris* under environmental and nutritional stress, evaluating the functional morphology of the midgut gland. Fernández-Gimenez (2002, 95) studied the effects of vitamin A to *P. muelleri* 's feed. Shrimp fed a vitamin free diet, 40 and 100 mg vitamin A kg^{-1}, manifested severe changes including shrinkage of cells and desquamation of apical borders, cellular death determinants, tissue disorganization and cellular lysis. Individuals fed 140 mg vit. A kg^{-1} showed less cellular damage. Some disorganized zones could be seen with hypertrophy and hyperplasia; however abundant secretion was identified in the lumen. The midgut gland of shrimps fed with 180 mg vit. A kg^{-1} was structurally normal and every cellular type could be identified in the tubules with their characteristic brush border (Figure 1). Over 180 mg kg^{-1} shrimp showed glands with an evident tissue disorganization, cellular retraction, hypertrophy and cellular lysis. The best well conserved structure of the gland was described for individuals given 160 mg vitamin A, in these cases some hypertrophic cells and retraction in a less percentage were found.

Fernández-Gimenez (2002, 86) also studied the effects of different levels of dietary vitamin A on the cytology of *A. longinaris'* midgut gland. Shrimp given between 0 and 100 mg vitamin A kg^{-1} diet, showed several histological changes, including hypertrophy and hyperplasia of tubular epithelium, folding of basal membrane and piknotic nucleus (Figures 2 and 3).

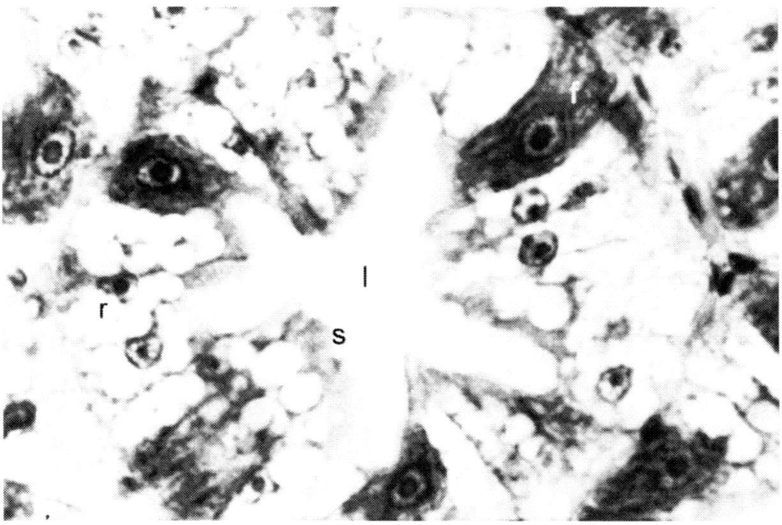

Figure 1. Midgut gland of *Pleoticus muelleri* fed a diet with 180 mg vitamin A kg^{-1}: transverse section through a tubule, showing all cellular types and brush border (similar to the found in wild shrimps). f, F cell; l, lumen; r, R cell; s, brush border (450X).

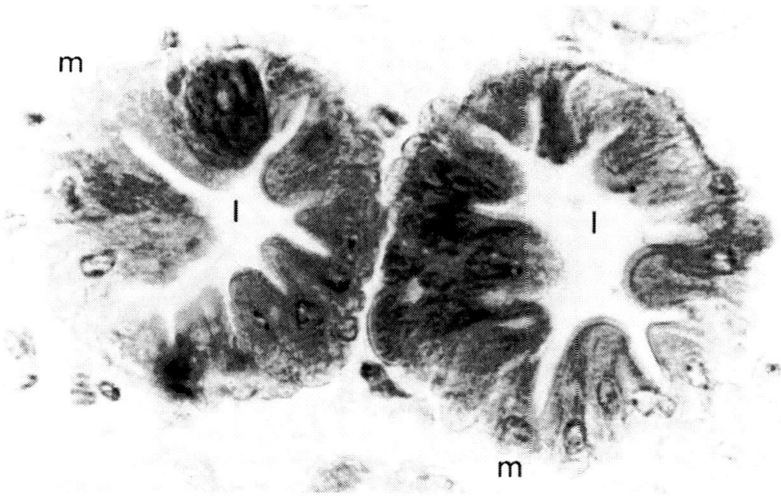

Figure 2. Midgut gland of *Artemesia longinaris* fed a diet deficient in vitamin A, showing basal membrane folding and cellular retraction. l, lumen; m, basal membrane folding. (450X).

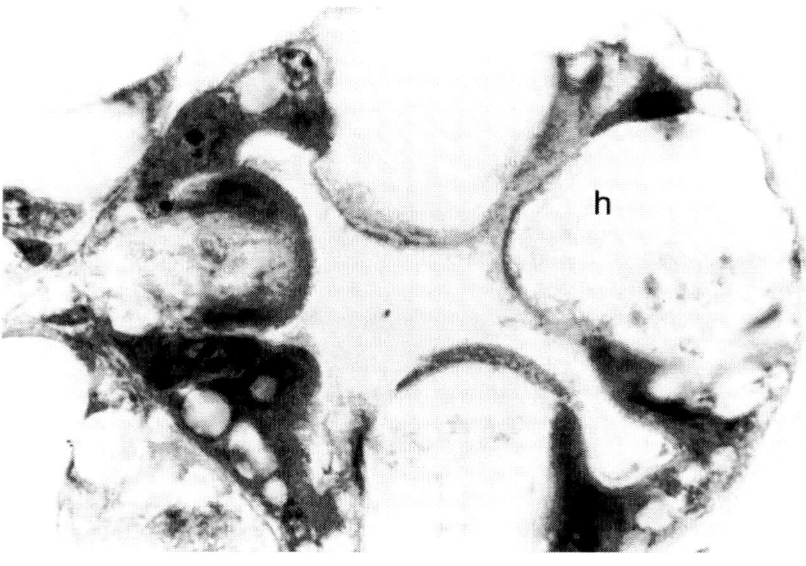

Figure 3. Midgut gland of *Artemesia longinaris* fed a diet deficient in vitamin A, detail of a tubule showing severe hypertrophy and cellular lysis. h, cellular hypertrophy. (450X).

In addition, individuals had given 100 mg to 200 mg vit. A kg^{-1} diet presented hypervacuolization, desquamation of apical border's cells, tissue disorganization, cellular lysis and blurred intercellular spaces. Haemocytic infiltration and some nodules were also identified. Individuals fed diets containing 300 mg vit. A kg^{-1} showed slight damage of the midgut gland.

The use of carotenoids as pigments in aquaculture feed is well documented. These pigments seem to have many physiological functions that include a role as antioxidant and provitamin A (Ponce-Palafox et al. 2006, 157). Crustaceans cannot synthesize carotenoids de novo but alter dietary carotenoids by oxidation and deposit them in their tissues. However, the effects of dietary carotenoid on shrimp physiological performance are varied, and the efficiency of carotenoid uptake from feed and deposition in shrimp tissues is poorly defined (Wade et al. 2015, 2).

In crustaceans, carotenoids stimulate the immune system, increase stress tolerance, serve as a source of vitamin A, and enhance embryonic development (Liñán-Cabello et al. 2002, 299). These nutrients are associated with the reproductive capacity of shrimp because they might contribute to cellular immuno-protection at critical stages, such as oocyte differentiation, that pose a high potential for free radical production (Liñán-Cabello et al. 2003, 386).

The use of bioactive substances such as nutritional additives to improve cultured shrimp yields is receiving increased attention and there are efforts to define the biological function of carotenoids as a dietary supplement. Díaz et al. (2011, 2) compared the growth and survival of juvenile shrimp *P. muelleri* fed diets containing different levels of carotenoids (50 and 100mg astaxanthin kg^{-1} diet; 50 and 100mg β-carotene kg^{-1} diet). These authors not founded evidence that support a possible influence of β-carotene or astaxanthin on growth or survival.

Recently, Wade et al. (2015, 4) evaluated the effects of a range of dietary astaxanthin inclusion levels (0, 25, 50 or 100 mg kg^{-1}) on growth, survival and carotenoid digestibility in shrimp *P. monodon*. The authors evidenced that shrimp fed carotenoid-free diets had significantly reduced colour and growth than those fed carotenoids, but survival was unaffected and carotenoid digestibility improved as dietary carotenoid levels increased. This study demonstrates that a total dietary carotenoid intake of between 25 and 50 mg kg^{-1} astaxanthin is required for normal shrimp growth and health in *P. monodon*.

VITAMIN D

Vitamin D_2 or ergocalciferol and vitamin D_3 cholecalciferol have vitamin D activity. Vitamin D_3 -also called 7-dehydrochelosterol- is formed in most animal tissues by the rupture of one of the ring bonds of 7-dehydrocholesterol by ultraviolet radiation (Raj 1987, 27). It is a steroid hormone and is well-known in vertebrates for maintaining calcium and phosphorus homeostasis and promoting bone mineralization. It modulates reproductive processes, may have a beneficial effect on insulin action by stimulating the expression of insulin receptors and thereby enhancing insulin responsiveness for glucose transport and has a modulating effect on the immune system. This vitamin does not function as a cofactor in any enzyme system (Holick 2007, 272; Whitaker 1994, 353).

There is no real indication that crustaceans regulate serum calcium levels in an analogous fashion to that found in vertebrates. Although the mineralized cuticle serves admirably as both a skeleton and a protective integument, its rigid structure presents problems to crustacean's growth. Crustaceans solved this difficulty with the molting. Some authors suggested that the resorptive processes occurring in the molting appear to be primarily for the conservation of organic constituents, not calcium. However, the existence in many crustaceans of several mineral storage mechanisms, such as gastroliths, seems to indicate a kind of regulation and thus a possible regulatory role of vitamin D. Gastroliths are paired mineralized structures formed between the epidermal and cuticular lining of the stomach and have a typical cuticular structure. Several problems arise in attempt to assign gastroliths and mineral regulation to a vitamin D system; gastroliths are not found in all crustaceans and their presence is not restricted to species inhabiting a specific environment related to calcium availability. They appear superfluous, particularly in marine species, which presumably can fulfill their calcium requirements from the surrounding medium. Most evidence indicates the molt and the formation of the gastroliths are both regulated by an ecdysteroid hormone, also gastroliths stores contain only a small portion both of the calcium that is resorbed from the integument before molting and of the calcium required following the molt (Conklin 1981, 150).

Deficient dietary vitamin D causes poor growth, high mortality, reduced appetite and darkening of midgut gland in *M. japonicus* and *L. vannamei* (Kanazawa 1984, 127; He et al. 1992, 181). Shiau and Hwang (1994, 2448) estimated a requirement of 0.2 mg Vit. D_3 kg^{-1} diet for *P. monodon* and Reddy et al. (1999, 269) observed that individuals of the same species fed vitamin D deficient feed grew poorly.

To determine the effects of vitamin D in growth and survival of *A. longinaris,* two trials were conducted with different levels of this nutrient in semi purified diets (0; 0.200; 0.375; 0.750 and 1mg kg^{-1}). The results of this study indicated that the dietary D requirements of this species, was around 0.375 and 0.750 mg kg^{-1} diet (Pereira 2011, 43), these levels are significantly higher than those found in *P. monodon* (Shiau and Hwang 1994, 2448). It is difficult to compare the requirements of different penaeoid shrimp species due to the different experimental conditions; for example, Shiau and Hwang (1994, 2446) carried out their investigations employing juvenile stages of *P. monodom*, meanwhile adults were used by Pereira (2011, 10).

Also, Pereira (2011, 24) studied the histology of midgut gland of *A. longinaris* as responses to dietary vitamin D_3, and the author observed that shrimp fed 0.375 and 0.75 mg vitamin D kg^{-1} seemed to conserve the gland as well as the ones given a complete diet (fishmeal, soybean meal and squid protein). Shrimp fed 0; 0.2 and 1 mg vitamin D kg^{-1} diet, showed several cytological damages as well as, an increment of tubular lumen, basal membrane folding, loss of brush border and cellular necrosis. The ones given 1 mg vitamin D kg^{-1} diet presented foamy cells.

Certain physiological processes in organisms are regulated via formation (phosphorylation) or rupture (dephosphorylation) of phosphate esters. Alkaline phosphatases (EC 3.1.3.1) are ubiquitous metalloenzymes located in the cell membrane involved in several essential functions in mammals. The knowledge of the structure and function of alkaline phosphatase has increased greatly in recent years and its activity determination is often used in clinical and ecotoxicological studies, abnormal levels in serum point out the existence of degenerative bone disease or liver damage (Nagata et al. 1989, 179).

Alkaline phosphatase has been studied and characterized in several invertebrates, Vijayavel and Balasubramanian (2006, 141) suggested that phosphatases play major roles in the molting physiology of many crustaceans. In the green crab *S. serrata,* this enzyme is important in the absorption of phosphate and calcium from seawater and for the integument formation (Yong-Doo et al. 2001, 766); however there are few studies in invertebrates which link the alkaline phosphatase activity and levels of vitamins in the diet.

Zhang et al. (2007, 53) investigated the interaction between vitamins A and D and alkaline phosphatase activity in Pacific abalone *Haliotis discus hannai*. The authors observed that dietary vitamin A generally increased the enzyme activity in viscera except the excessive supplement which decreased phosphatase activity and dietary vitamin D significantly increased enzyme activity. On other hand, in crustaceans, Shiau and Hwang (1994, 2449) suggested that alkaline phosphatase may be involved in the metabolism of calcium and phosphorus in shrimp, .they reported that absence and levels above 0.1 mg vitamin D kg^{-1} in shrimp *P. monodon* diet, correlates with a reduced activity of this enzyme.

Alkaline phosphatase activity was evidenced by Pereira et al. (2013, 7) in haemocytes and midgut gland of shrimp *A. longinaris*. The purpose of this study was to evaluate the enzyme activity in relationship with different doses of vitamin D_3 (0; 0.2; 0.375; 0.75 and 1 mg vitamin D_3 kg^{-1}) in feed and to estimate its potential use as biomarker for nutritional stress. The results demonstrated higher enzymatic activity values in haemocytes than in midgut glands. The highest enzyme activity in haemocytes was observed for shrimp fed without vitamin D (1.235 abs min^{-1} mg $protein^{-1}$), however, in midgut gland, the activity varied from 0.141 to 0.297 abs min^{-1} mg $protein^{-1}$, with the highest values on 0 and 1 mg vitamin D kg^{-1} feed. Histological analysis of the midgut gland confirmed a good health of the shrimp fed 0.375 and 0.750 mg vitamin D_3 kg^{-1} feed and were used as optimal values for determining enzymatic activity. The results indicated that alkaline phosphatase activity in *A. longinaris* was influenced by dietary vitamin D and may be used as a biomonitor of nutritional stress.

VITAMIN E

Vitamin E is the terminology for a group of lipid-soluble compounds, α, β, γ and δ-tocopherols and tocotrienols. The vitamin E compound with highest biological activity is α-tocopherol; its main function is to protect unsaturated lipids in living tissues against free-radical-mediated oxidation. As a fat soluble vitamin, it is the most effective chain-breaking, lipid soluble antioxidant in biological membranes, where it contributes to membrane stability; it protects critical structures against damage from oxygen free radicals and reactive products of lipid peroxidation (Hsu and Shiau 1999,169; Lee and Shiau 2004, 481).

Several studies showed that dietary vitamin E was required by the water flea *Daphnia magna* (Viehoever and Cohen 1938, 297) and *Moina macrocopa* (Conklin and Provasoli 1997, 343). Kanazawa (1984, 127) found that an addition of vitamin E to feed results in survival improvement of larval *M. japonicus* and He et al. (1992, 182) reported that *L. vannamei* showed significantly lower survival and weight gain when fed vitamin E free diet for 8 weeks. Reddy et al. (1999, 269) found significantly depressed growth and poor appetite in shrimp *P. monodon* fed a vitamin E deficient diet.

Two trials to evaluate the growth and survival of shrimp *P. muelleri* fed semipurified diets containing different levels of vitamin E have been carried out (Fernández-Gimenez et al. 2004, 1173). In the first experiment the author did not find any relation between weight gain and survival to the dietary level of vitamin E (0; 100; 600 and 1500 mg kg^{-1} diet), however in the second trial, found a positive correlation between the amount of vitamin E in diets and growth (1250; 1500 and 1750 mg kg^{-1} diet) and the lowest weight gain was observed in shrimp fed with 2000 mg kg^{-1} diet. The best growth was obtained when diet was supplemented with 1750 mg kg^{-1} feed. Additionally, the histology of midgut gland of shrimp fed 1750mg vitamin E kg^{-1}, showed a well preserved cytology, and every cellular type could be recognized as well as the brush border and secretion inside the tubules indicating cellular activity. The animals of the others treatments presented cellular retraction, picnotic nuclei or loss of nucleus, desquamation of the apical zone and degradation of the basal membrane and some cases of retraction and hypertrophy were described. A similar requirement (1500 mg kg^{-1}) was obtained for shrimp *A. longinaris* previously (Fernández-Gimenez and Fenucci 2002, 3).

The study of *A. longinaris*' midgut gland fed with diets containing less than 1000 mg vitamin E kg^{-1} showed a severe deterioration of the gland with connective tissue and haemocyte infiltration, some cases of encapsulation, cell retraction, hypertrophy and hyperplasia, lysis and necrosis. Shrimp given over than 1500 mg vitamin E, presented reduced intertubular spaces, brush border and abundant secretion of the lumen, indicating that both structure and functionality of the gland were well preserved, there were scarce signs of cellular retraction and lysis (Fernández-Gimenez 2002, 96).

The requirements of two Argentine shrimp species are significantly higher than the dosages described for other Decapoda. For example, an optimum requirement of 99 mg vitamin E kg^{-1} diet was determined for *L. vannamei* (He et al. 1992, 181; He and Lawrence 1993, 249) and for maximal growth of *P. monodon* (Hsu and Shiau 1999, 170) the level of vitamin E required was 85-95 mg vitamin E kg^{-1}. Aquatic animals have high levels of unsaturated fatty acids to maintain cell membrane fluidity especially at low temperatures. This may explain why *P. muelleri* and *A. longinaris*, as species that live in cold waters of South West Atlantic, need an important amount of Vitamin E in the diet, for these, it is assumed that this compound plays an important role in both Argentine penaeoid species. Furthermore, it is difficult to compare the requirements of the different penaeoid shrimp due to the different experimental conditions; Hsu and Shiau (1999, 169) and He et al. (1992, 178) carried out their investigations employing larval states of *P. monodon* and *L. vannamei*, respectively, meanwhile adults of *P. muelleri* and *A. longinaris* were used by Fernández-Gimenez (2002, 62).

When working with vitamin E, it is also important to consider the lipid concentration in feed because this compound's level increases when polyunsaturated fatty acid's level increases in feed. Considering previous investigations, it is possible that the differences in the estimated requirement may vary upon feed composition. The incorporation of good quality oils to feeds protects the lipids during the manufacture and storing and apparently reduces the vitamin E requirement in the trout (Hung et al. 1981, 653). The level and state of oxidation of polyunsaturated lipids in food, as well as the presence of the other antioxidants and selenium, affect the dietary level of vitamin E required by fish (Roem et al. 1990, 155).

He and Lawrence (1993, 247) fed shrimp *L. vannamei* with feed supplemented with fatty acids and corn oil without vitamin E, meanwhile the vitamin C (a great antioxidant) was present. On other hand, Fernández-Gimenez et al. (2004, 1173) used refined fish oil completely free of vitamins and antioxidants on formulated feed for shrimp *P. muelleri*, besides in these foods the only protection against lipid peroxidation was supplied by vitamin E supplementation.

In the Indian prawn *Fenneropenaeus indicus* was determined the effect of 100 and 300 mg vitamin E kg^{-1} diet along with two levels of lipid (9 and 14%) and their interaction on growth performance (Ouraji et al. 2011, 904) and consequently evaluated the fatty acid composition and lipid stability on its muscle tissue during frozen storage. Results showed that an increasing concentration of vitamin E from 100 to 300 mg kg^{-1} in diet reduced the rate of lipid oxidation in the muscle tissue during frozen storage and caused higher Highly Unsaturated Fatty Acids (HUFA) retention in muscle of shrimp fed high lipid level diets. However there was no significant interaction between dietary vitamin E and lipid levels on growth performance. They also described malonaldehyde as a secondary oxidation product of polyunsaturated fatty acids and its analysis is one of the most popular and commonly used methods for tissue peroxidation determination and an increasing dietary vitamin E level significantly lowered the malonaldehyde values in shrimp muscle by lowering the rate of lipid peroxidation during frozen storage.

Oxygen is an essential compound of living organisms and the generation of reactive oxygen species (ROS) such as superoxide anion, hydrogen peroxide, hydroxyl radicals, and singlet oxygen, commonly referred as prooxidants, is inevitable in aerobic metabolism of the body. ROS cause lipid oxidation, protein oxidation, DNA strand break and base modification, and modulation of gene expression which finally lead to several pathological states including apoptosis (Lee et al. 2004, 24).

Therefore, in living systems an equilibrium exists between the rate of production and the rate of breakdown of ROS. Under normal physiological states, harmful effects of ROS are effectively neutralized by antioxidant defense system of organisms, which generally comprises of both enzymes and small antioxidant molecules. However an alteration of the prooxidant and antioxidant equilibrium in favors of the former, results in oxidative stress (Dandapat et al. 2000, 102).

Aquatic pollution is a major contributor to oxidative stress in fish and crustaceans. These pollutants are taken up by aquatic organisms from sediments, suspended particles matter with toxic properties and food sources (Barim and Karatepe 2010, 141). A significant role of antioxidant defenses is recognized in protecting aquatic animals from oxidative stress which resulted due to their exposure to various agents such as xenobiotics, nutritional imbalance and hypoxia. Among the various antioxidants, vitamin E is unique and deserves special attention for its important role in maintaining the quality of aquaculture products. It also provides additional health protection through its immuno-stimulant property (Dandapat et al. 2000, 102). In crustaceans the midgut gland, which is rich in lipids, contains fat soluble vitamins and regulates the body metabolism. It is recommended as an environmental indicator of water pollution (Borkovic et al. 2008, 127) and as biomarker of oxidative stress (Valavanidis et al. 2006, 182). Barim and Karatepe (2010, 141), determined that vitamins A, E, C and β-carotene in midgut gland and muscle tissues of crayfish *Astacus leptodactylus* were significantly higher at unpolluted water, than polluted, providing the evidence that vitamins A, E, C and β-carotene are sensitive indicators of aquatic pollution.

Dandapat et al. (2000, 110) studied the effect of vitamin E in diets (200, 400 and 600 mg kg^{-1}) on lipid peroxidation in gills and midgut gland of the giant river prawn *Macrobrachium rosembergii.* The authors suggest that vitamin E is capable of reducing lipid peroxidation and can modulate the antioxidant defense system in gills and midgut gland, nevertheless, the response is highly specific. It is further observed that highest dose of vitamin E (600 mg kg^{-1}) could not render much additional protection in both the tissues.

On the other hand, Díaz et al. (2004, 3192) evaluated the free radical scavenging properties of *P. muelleri's* midgut gland extracts by electron paramagnetic spin resonance spectrometry methods (EPR) against the stable 1,1-diphenyl-2-picrylhydrazyl (DPPH) radical. This study was carried out to characterize different physiological stages of the shrimp under environmental and nutritional stress. Feeding trials were carried out on juveniles and feeds with different concentrations of vitamins A and E were formulated. The control groups were fed with fresh squid mantle and with a vitamin free diet. For all treatments, the extracts exhibited strong DPPH radical scavenging activity, suggesting that the tissue is a powerful natural antioxidant. Individuals fed with different concentrations of vitamin E showed the strongest effect of the DPPH radicals, reducing the DPPH radicals to 50%, after an incubation period of 3 min. In contrast, the extracts of control shrimp, fed with squid mantle, had the weakest antioxidant activity. These authors suggest that loss of individual antioxidants and/or generation of oxidation products from them can be measured as an index of oxidative stress and the presence of vitamins A or E in feed can provide immediate protection against free radicals.

VITAMIN K

Vitamin K is associated to normal blood coagulation and calcium deposition in vertebrates. It is required for carboxylation of glutamate residues to Gamma-carboxyglutamate residues in many proteins, which are necessary for normal blood coagulation. Philloquinone (vitamin K_1) is synthesized by plants and algae, whereas the menaquinone family (MK-*n*) (vitamin K_2) includes products of bacterial biosynthesis. Vitamin K_2 is a cofactor for a number of bacterial steroidal-*P*-dehydrogenases and it is involved in the blood clotting process in animals and the production of some of the intermediates of metabolic pathways. Water soluble salt of the synthetic menadione is used in animal diets: fish feeds are commonly supplemented with menadione sodium bisulfite (Grahl-Madsen and Lie 1997, 269-270).

The requirement for vitamin K has been studied in several species of fishes and crustaceans. Its deficiency results in anemia and prolonged coagulation time in fish and it is highly effective in preventing the molinate-induced anemia in common carp (Kawatsu et al. 1989, 779). In crustaceans, vitamin K is required for *P. monodon* (Shiau and Liu 1994a, 278), *F. chinensis* (Shiau and Liu 1994b, 129), *P. muelleri* and *A. longinaris* (Sarasa 2010, 34).

Vitamins D and K are involved in calcium and phosphorus transport in vertebrates (Ceccaldi 1997, 267), although vitamin K has been poorly studied in crustaceans, it has been proposed to play the same role as in vertebrates. Moreover, Shiau and Liu (1994a, 280) found in the midgut gland of *P. monodon* evidence of a carboxylase and a protein precursor dependent on menadione and they proposed that calcium deposition in juvenile tiger prawn positively correlates with the levels of vitamin K and D in diet. They noticed higher weight gain in grass shrimp *P. monodon* fed diets containing 35 mg menadione kg^{-1} diet.

Shiau and Liu (1994b, 132) recorded high weight gain of individuals of *F. chinensis*, given diets supplemented with vitamin K. They obtained good weight increase for juvenile shrimp fed diets between 160 and 320 mg menadione kg^{-1}, determining an optimal value of 185 mg kg^{-1}. In contrast to these, no effect in weight gain and survival for *L. vannamei* fed a full vitamin supplemented diet and a feed without vitamin content was observed (He et al. 1992, 181). Sarasa (2010, 34) found no statistical differences in weight gain and survival for shrimp *P. muelleri* and *A. longinaris* fed a fully supplemented diet and 0, 160, 240 and 320 mg vitamin K_3 $diet^{-1}$.

Sarasa (2010, 23) observed histological alterations that indicated that a supplementation of menadione in the diet was necessary to maintain the normal structure of midgut gland of *P. muelleri*. Diets without vitamin K or with 160 and 320 mg vit. K kg^{-1} leaded to severe changes of the cytology of the gland, including cellular retraction, absence of brush border, hypertrophy, desquamation and necrosis. Shrimp fed 240 mg vitamin K kg^{-1} diet and a complete diet showed a normal functional morphology of midgut gland.

Studying the same diets on *A. longinaris*, Sarasa (2010, 22), found out that inadequate levels of menadione (160 and 320 mg vit. K $diet^{-1}$) and the absence of this compound, generated cellular hyperplasia and hypertrophy, disorganization of the midgut gland tissue and shrinkage of cells. The most preserved organs were described for individuals given complete diets or 240 mg vit. K $diet^{-1}$ supplemented ones. Also, Sarasa (2010, 24) observed an inverse relationship between alkaline phosphatase activity in midgut gland of *P. muelleri* and levels of vitamin K_3 in diet. In larvae of white shrimp *Litopenaeus setiferus* it was demonstrated that alkaline phosphatase activity is situated along the midgut gland, suggesting that absorption is widespread, however in juveniles the activity is restricted to the midgut gland and midgut region. Alkaline phosphatase activity in the digestive gland of decapods has been associated with metabolites transmembrane transport; tissues in which this enzyme activity is present are generally thought to function in active transport absorption (Lovett and Felder 1990, 166-167). Using histochemical methods, Monin and Rangneker (1974, 1-2) observed that midgut gland cells

·of crab *S. serrata*, showed a positive reaction for alkaline phosphatase only at the brush border. Physiological processes, as molting, also affect AP activity. Meyran and Craf (1986, 313) observed periodical changes in the enzyme activity in the posterior caeca of the amphipod *Orchestia cavimana*, using cytochemical techniques during the molting cycle; suggesting that this enzyme may be involved in calcium transport. Other vitamin had also a similar effect on alkaline phosphatase activity of midgut gland. Sarasa (2010, 34) observed that shrimp *P. muelleri* responded to different levels of vitamin K in feed; in this case the highest phosphatase alkaline activity was linked to several histological changes in the midgut gland.

CONCLUSION

Penaeoid shrimp, *A. longinaris* and *P. muelleri* are two important marine resources in the Argentine Sea, with commercial importance. Both species have seasonal and annual fluctuations in catches, it is, therefore, important to establish the feasibility of culturing them on commercial basis to provide continuous supplies of these species to the market. The success of culturing aquatic species resides in the inclusion of tools that are able to measure the interaction between food, environment and physiology of the organisms. Dietary vitamin requirements in crustaceans have been demonstrated with dietary trials for growth and survival; however, vitamin deficiencies and over fortifications may be manifested through changes in the midgut gland cells. This chapter provides a better understanding of relation between liposoluble vitamins in diet and metabolic, physiological and histological responses, which provides knowledge about biology of this species and the formulation of suitable feeds for shrimp (Table 1). As the midgut gland is the primary digestive organ of crustaceans, pathological changes of these cells result in a general impairment of the digestive function. The most notable changes reported for *P. muelleri* and *A. longinaris* fed with different levels of liposoluble vitamins, included cellular and nuclear retraction, desquamation or detachment of cells and damage of the basal lamina, and have been reported for other penaeid species, such as *P. monodon* and *F. indicus* (Reddy et al. 1999, 272). Biochemical studies have demonstrated the relationship between levels of vitamin in diet and enzyme activity, such as, alkaline phosphatases, and the important role of vitamins as antioxidants and immunomodulators.

Table 1. Optimum levels of dietary liposoluble vitamins for Argentine shrimp species

Species	Vitamins	Optimum levels (mg/kg)	Weight gain (%)
Artemesia longinaris	A	50	52.1
	D	0.375	35.6
	E	1500	71.0
	K	320	22.0
Pleoticus muelleri	A	160	73.0
	E	1750	65.0
	K	320	29.1

REFERENCES

Alava, Veronica, Kanazawa, Akio, Teshima, Shin-ichi, and Koshio, Shunsuke. 1993. "Effect of dietary vitamin A, E and C on gonadal maturation of *Penaeus japonicus." Nippon Suisan Gakkaishi,* 59: 1235-1241.

Al-Mohanna, Salem, and Nott, James. 1989. "Functional cytology of the hepatopancreas of *Penaeus semisulcatus* (*Crustacea: Decapoda*) during the moult cycle." *Marine Biology,* 101: 535-544.

Barim, Ozden, and Karatepe, Mustafa 2010. "The effects of pollution on the vitamins A, E, C, β-carotene contents and oxidative stress of the freshwater crayfish, *Astacus leptodactylus." Ecotoxicological and Environ mental Safety,* 73: 138-142.

Bearer-Rogers, Joyce, Dieffererbacher, Albrecht, and Holm, Jensen. 2001. "Lexicon of lipid nutrition (IUPAC Technical Report)." *Pure and Applied Chemistry,* 73: 685-744.

Borkovic, Slavica, Pavlovic, Sladjan, Kovacevic, Tijana, Stajn, Andras, Petrovic, Vojislav, and Saicic Zorica. 2008. "Antioxidant defense enzyme activities in hepatopancreas, gills, and muscle of Spiny cheek crayfish (*Orconectes limosus*) from the River Danube." *Comparative Biochemistry and Physiology,* 147C: 122-128.

Castell, John 1989. "Reference diet for crustaceans: principles of experimentation." Paper presented at Advances in Tropical Aquaculture, Aquacop, IFREMER, February 20-March 4.

Ceccaldi Hubert J. 1997. "Anatomy and physiology of the digestive system." In *Crustacean Nutrition*, edited by Lawrence R D'Abramo, Douglas E

Conklin and Dean M Akiyama, 261-290. Baton Rouge: World Aquaculture Society.

Combs Gerald F. 1998. The vitamins: Fundamental Aspects in Nutrition and Health. San Diego: Academic Press.

Conklin Douglas E., and Provasoli Luigi. 1997. "Nutritional requirements of the water flea *Moina macrocopa*." *Biological Bulletin,* 152: 337-350.

Conklin Douglas E. 1981. "The role of micronutrients in the biosynthesis of the crustacean exoesqueleton." Paper presented at the second international conference on aquaculture nutrition, Biochemical and Physiological Approaches to Shellfish Nutrition, Baton Rouge, November 19-21.

Conklin Douglas E. 1989. "Vitamin requirements of juvenile penaeid shrimp." Paper presented at Advances in Tropical Aquaculture, Aquacop, IFREMER, February 20-March 4.

Dandapat, Jagneshwar, Chainy, Gagan B. N., and Rao Janardhana J. 2000. "Dietary vitamin-E modulates antioxidant defense system in giant freshwater prawn, *Macrobrachium rosenbergii*." *Comparative Biochemistry and Physiology,* 127C: 101-115.

Díaz, Ana C., Fernández-Gimenez Analía V., Mendiara, Sara C., and Fenucci Jorge L. 2004. "Antioxidant activity in hepatopancreas of the shrimp (*P. muelleri*) by electron paramagnetic spin resonance spectrometry." *Journal of Agriculture and Food Chemistry*, 52: 3189-3193.

Díaz, Ana C., Velurtas, Susana M., Fernández-Gimenez, Analía V., Mendiara, Sara N., and Fenucci, Jorge L. 2011. "Carotenoids in Integument, Muscle and Midgut gland of Red shrimp *Pleoticus muelleri* Bate, 1888 (Crustacea, Penaeoidea) fed carotenoid-supplemented diets." *Israeli Journal of Aquaculture-Bamidgeh,* 63:625-632.

Fenucci, Jorge L., Díaz, Ana C., and Fernández-Gimenez, Analía V. 2009. "A Review on the Status of Protein Nutrition of Argentine Penaeoid shrimp: comparisons and contrasts within the Penaeidae." In *The Rising Tide*, edited by Craig L. Browdy, Darryl F. Jory, 148-160. Baton Rouge: World Aquaculture Society.

Fernández-Gimenez, Analía V., and Fenucci Jorge L. 2002. "Vitamin E requirement of the prawn *Artemesia longinaris* (*Decapoda, Penaeidae*)." In *Modern Aproaches to the study of Crustacea*, edited by Elva Escobar-Briones, Fernando Alvarez, 85-90. New York: Kluwer Academia/Plenum Publishers.

Fernández-Gimenez, Analía V., Díaz, Ana C., Velurtas, Susana M., Petriella, Ana M., and Fenucci, Jorge L. 2008. "Effects of different dietary vitamin A levels in the red shrimp *Pleoticus muelleri* (Bate, 1888) (Decapoda,

Solenoceridae)." *Revista de Biología Marina y Oceanografía,* 43: 483-490.

Fernández-Gimenez, Analía V., Fenucci, Jorge L. and Petriella, Ana M. 2004. "The effect of vitamin E on growth, survival and hepatopancreas structure of the Argentine red shrimp *Pleoticus muelleri* Bate (Crustacea, Penaeoidea)." *Aquaculture Research,* 35:1172-1178.

Fernández-Gimenez, Analía V., Haran, Nora S., Pereira, Nair A., Sarasa, María V., and Fenucci, Jorge L. 2010. "Capacidad inmunoestimulante de la vitamina D en el camarón argentino *Artemesia longinaris*." Paper presented at III Foro Iberoamericano Recursos Marinos y Acuicultura, Hermosillo, México, November 10-12.

Fernández-Gimenez, Analía V. 2002. "*Requerimiento de vitaminas liposolubles A y E en la dieta del camarón Artemesia longinaris Bate, 1888 y el langostino Pleoticus muelleri (Bate, 1888)*." Ph.D. diss., Buenos Aires University.

Fisher, Robert, Kon, Stanislaw K., and Thompson Stan Y. 1952. "Vitamin A and carotenoids in certain invertebrates. I. Marine Crustacean." *Journal of the Marine Biological Association UK (JMBA),* 31: 229-258.

Fisher Robert, Kon Stanislaw K., and Thompson, Stan Y. 1957. "Vitamin A and carotenoids in certain invertebrates, VI Marine Crustacean." *Journal of the Marine Biological Association UK (JMBA),* 36: 501-507.

Grahl-Madsen, Elisabeth, and Lie, Øyvind. 1997. "Effects of different levels of vitamin K in diets for cod (*Gadus morhua*)." *Aquaculture,* 151: 269-274.

Halver, John E. 1957. "Nutrition of Salmonid Fishes, III Water Soluble vitamin requirements of Chinook salmon." *Journal of Nutrition,* 62: 225-243.

He, Haiqui, and Lawrence, Addison L. 1993. "Vitamin E requirement of *Penaeus vannamei*." *Aquaculture,* 118: 245-255.

He, Haiqui, Lawrence, Addison L., and Liu, Ruiyu. 1992. "Evaluation of dietary essentiality of fat-soluble vitamins, A, D, E and K for penaeid shrimp (*Penaeus vannamei*)." *Aquaculture,* 103: 177-185.

Holick, Michael F. 2007. "Vitamin D deficiency." *The New England Journal of Medicine,* 357: 266-281.

Hsu, Tsai-Shen, and Shiau, Shi-Yen. 1999. "Tissue Storage of Vitamin E in Grass Shrimp *Penaeus monodon* fed dietary DL-α-tocopherol Acetate." *Fisheries Science*, 65: 169-170.

Hung, Silas S. Cho, Chen Y., and Slinger, Stanley J. 1981. "Effect of oxidized fish oil, dl-α tocopheryl acetate and ethoxyquin supplementation on the vitamin E nutrition of rainbow trout (*Salmo gairdneri*) fed practical diets." *Journal of Nutrition,* 3: 648-657.

Kanazawa, Akio. 1984. "Nutrition of penaeid prawn and shrimp." Paper presented at the First International Conference on the Culture of Penaeid Prawns/Shrimps, Southeast Asian Fisheries Development Center, Iloilo City, Philippines, December 4-7.

Kawatsu, Hiroshi, Kondo, Kengo, and Kubono, Kazushige. 1989. "Hemostatic disorder in common carp induced by exposure to the herbicide molinate." *Bulletin of the Japanese Society for the Science of Fisheries,* 55: 779-783.

Lee, Joo, Koo, Nan, and Min, David B. 2004. "Reactive Oxigen Species, Aging and Antioxidative Nutraceuticals." *Comprehensive Reviews in Food Science and Food,* 3: 21-33.

Lee, Min-Hsien, and Shiau, Shi-Yen. 2004. "Vitamin E requirements of juvenile grass *Penaeus monodon* and effects on non-specific immune responses." *Fish and Shellfish Immunology,* 16: 475-485.

Liang, Mengqing and Ji, Wang. 1998. "Study on nutritional requirement of vitamin A for Chinese prawn *Penaeus chinensis* larva." *Marine Fisheries Resources,* 19: 86-90.

Liñán-Cabello, Marco A., Paniagua, Michel Jesus, and Hopkins, Paul M. 2002. "Bioactives roles of carotenoids and retinoids in crustaceans." *Aquaculture Nutrition,* 8: 299-309.

Liñán-Cabello, Marco A., Paniagua, Michel Jesus, and Zenteno-Savín, Tania. 2003. "Carotenoids and retinal levels in captive and wild shrimp, *Litopenaeus vannamei*." *Aquaculture Nutrition,* 9: 383-389.

Lovett, Donald L., and Felder, Darryl L. 1990. "Ontogenetic changes in enzyme distribution and midgut function in developmental stages of *Penaeus setiferus* (*Crustacea, Decapoda, Penaeidae*)." *Biological Bulletin,* 178: 160-174.

Mengqing, Liang, Wenjuan, Jobgen, Qing, Carl, and Wgcjsun Jialin. 2004. "The effect of vitamin A supplementation in broodstock feed on reproductive performance and larval quality in *Penaeus chinensis*." *Aquaculture Nutrition,* 10: 295-300.

Meyran, Jean-Claude, and Craft, François. 1986. "Ultrahistochemical localization of Na+K+ ATPase, Ca2+ ATPase and alkaline phosphatase activity in a calcium-transporting epithelium of a crustacean during moulting." *Histochemistry,* 85: 313-320.

Monin, Mansoor A., and Rangneker, Padgaonkar V. 1974. "Histochemical localization of acid and alkaline phosphatases and glucose 6-Phosphatase of the hepatopancreas of the crab *Scylla serrata* (Forskål)." *Journal of Experimental Marine Biology and Ecology,* 14: 1-16.

Nagata, Akiya, Komoda, Tsugikazu, and Sakagishi, Yoshikatsu. 1989. "Relationship between the Uptake of Calcium or Phosphorus and Alkaline Phosphatase Activity Induced by Certain Modulators in Rat Organs." *Calcified tissue international,* 45: 173-181.

National Research Council 2011. *Nutrient Requirements of Fish and Shrimp.* Washington DC: Academic Press.

Ouraji, Hossein, Abedian Kenari, Abdolmohammad M., Shabanpour, Bahareh, Shabani, Ali, Sodagar, Mohammad, Jafarpour, Seyed A., and Ebrahimi, Ghaffar H. 2011. "Growth, survival, and fatty acid composition of Indian white shrimp *Fenneropenaeus indicus* (Milne Edwards) fed diets containing different levels of vitamin E and lipid." *Aquaculture International,* 19: 903-916.

Pereira, Nair A., Fernández-Gimenez, Analia V., Haran, Nora S., and Fenucci, Jorge L. 2013. "Alkaline Phosphatase in Shrimp *Artemesia longinaris*: response to Feed." *International Journal of Aquatic Science,* 4: 3-12.

Pereira, Nair A. 2011. "*Vitamina D_3 en la dieta del langostino Pleoticus muelleri y el camarón Artemesia longinaris en cultivo.*" Bachelor diss., Mar del Plata University.

Ponce-Palafox, Jesús T., Arredondo-Figueroa, José L., and Vernon-Carter, Enrique J. 2006. "Carotenoids from plants used in diets for the culture of the Pacific white shrimp (*litopenaeus vannamei*)." *Revista Mexicana de Ingenieria Química,* 5: 157-165.

Raj, Paul R. 1987. "Vitamin Requirements of Finfish and Prawns." Paper presented at the Summer Institute Recent Advances in Fishfish and Shellfish Nutrition, Cochin, Japan, May 11-30.

Reddy, Harish R. V., Rai, Ashwin, and Annappaswamy, Setty. 1999. "Evaluation of the dietary essentiality of vitamins for *Penaeus monodon.*" *Aquaculture Nutrition,* 5: 267-275.

Rodriguez Souza, Juan C., Sekine, Shintaro, Susuki, Satoko, Shima, Yasushiro, Stüssmann, Carlos A., and Takashima, Fumio. 1996. "Usefulness of histological criteria for assesing the adequacy of diets for *Palinurus japonicus* phyllosoma larvae." *Aquaculture Nutrition,* 2: 133-140.

Roem, Andries, Kohler, Christopher C., and Stickney, Robert R. 1990. "Vitamin E requirements of the blue tilapia, *Oreochromis aureus* (*Steindachner*) in relation to dietary lipid level." *Aquaculture,* 87: 155-165.

Sarasa, María V. 2010. "*Vitamina K en la dieta del langostino Pleoticus muelleri y el camarón Artemesia longinaris en cultivo.*" Bachelor diss., Mar del Plata University.

Shiau, Shi-Yen, and Hwang, Jean-Yu. 1994. "The Dietary Requirement of Juvenile Grass Shrimp (*Penaeus monodon*) for Vitamin D." *Journal of Nutrition,* 124: 2445-2450.

Shiau, Shi-Yen, and Liu, Jia-Shiun. 1994a. "Quantifying the vitamin K requirement of juvenile marine shrimp (*Penaeus monodon*) with menadione." *Journal of Nutrition,* 124: 277-282.

Shiau, Shi-Yen, and Liu, Jia-Shiun. 1994b. "Estimation of the dietary vitamin K requirement of juvenile *Penaeus chinensis* using menadione." *Aquaculture,* 126: 129-135.

Tacon, Albert G. J., Metian, Marc, and Hasan, Mohammad R. 2009. *Feed ingredients and fertilizer for farmed aquatic animals, Sources and composition.* Rome: Food and Agriculture Organization of the United Nations.

Tacon, Albert G. J. 1991. "Vitamin nutrition in shrimp and fish." Paper presented at the Aquaculture Feed Processing and Nutrition Workshop, American Soybean Association, Singapore, September 19-25.

Valavanidis, Athanasios, Vlahogianni, Thomais, Dassenakis Manos, and Scoullos, Michael. 2006. "Molecular biomarkers of oxidative stress in aquatic organisms in relation to toxic environmental pollutants." *Ecotoxicological Environmental Safety,* 64: 178-189.

Viehoever, Arno and Cohen Isadore. 1938. "The response of *Daphnia magna* to vitamin E." *American Journal of Pharmacy,* 110: 297-315.

Vijayavel, Kannappan and Balasubramanian, Mohan P. 2006. "Fluctuations of biochemical constituent and marker enzymes as a consequence of naphthalene toxicity in the edible estuarine crab, *Scylla serrata.*" *Ecotoxicological Environmental Safety,* 63: 141-147.

Wade Nicholas M., Cheers Sue, Bourne Nicholas, Irvin Simon, Blyth David and Glencross Brett D. 2015. "Dietary astaxanthin levels affect colour, growth, carotenoid digestibility and the accumulation of specific carotenoid esters in the Giant Tiger Shrimp, *Penaeus monodon.*" *Aquaculture Research,* 1-12. doi:10.1111/are.12888.

Whitaker, John R. 1994. *Principles of Enzymology for the Food Sciences.* New York: Marcel Dekker.

Wolfe, Douglas A. and Cornwell, David G. 1965. "Composition and tissue distribution of carotenoids in crayfish." *Comparative Biochemistry and Physiology,* 16: 205-213.

Yong-Doo, Park, Yi, Yang, Qing-Xi, Chen, Hai-Ning, Lin, Quiang, Liu, and Hai-Meng, Zhou. 2001. "Kinetics of complexing activation by the magnesium ions on the activity of green crab (*Scylla serrata*) alkaline phosphatase." *Biochemistry Cell Biology,* 79: 765-772.

Zhang, Wenbing, Mai, Kangsen, Xu, Wei, Tan, Beiping, Ai, Qinghui, Liufu, Zhiguo, Ma, Hongming, Wang, Xiaojie. 2007. "Interaction between vitamins A and D on growth and metabolic responses of abalone *Haliotis discus hannai*, Ino." *Journal of Shellfish Research,* 26: 51-58.

Reviewed by: Dr. Juan Carlos Mallo, Profesor/Investigador, Universidad Tecnológica Nacional, Mar del Plata, Argentina. E-mail: jcmallo@mdp.edu.ar.

In: Crustaceans
Editor: Victor Alvarado

ISBN: 978-1-63484-890-9

Chapter 4

CRUSTACEANS FROM TEMPERATE SANDY BEACHES: A POSSIBLE ROUTE FOR MANAGEMENT AND CONSERVATION STRATEGIES

***Sílvia C. Gonçalves*[1,2,*]**

[1]MARE – Marine and Environmental Sciences Centre, ESTM, Polytechnic Institute of Leiria, Peniche, Portugal

[2]MARE – Marine and Environmental Sciences Centre, Department of Life Sciences, Faculty of Sciences and Technology, University of Coimbra, Coimbra, Portugal

ABSTRACT

Crustaceans are important components of the macrofaunal communities of sandy beaches worldwide and may present a key role in the structure and function of these ecosystems in certain conditions. These animals represent the dominant faunal group in exposed beaches, where some crustacean species form abundant populations, especially in the supralittoral zone. In temperate latitudes, Talitrid amphipods, commonly known as sand hoppers, and Tylid isopods are often abundant elements in exposed sandy beaches although, due to their nature, Talitrids

* Corresponding author: ESTM – School of Tourism and Maritime Technology, Polytechnic Institute of Leiria, Campus 4, Santuário Na. Sra. dos Remédios, 2520-641 Peniche, Portugal. Tel: +351 262 783 607. Fax: +351 262 783 088, scgoncalves@ipleiria.pt.

are the most conspicuous ones. Despite their clear ecological relevance as interface systems between the sea and land, sandy beaches also constitute unique and irreplaceable habitats for most of these Crustacean species. However, the intensification of human activities in coastal areas observed in the last decades, as well as the global climatic changes and coastal erosion occurring nowadays, have introduced several detrimental impacts threatening these ecosystems and their biota. In this commentary, the relevance of using Talitrid and Tylid crustaceans as ecological tools in management and conservation strategies for temperate sandy beaches is discussed. Also, the benefits of these strategies for the conservation of the unique habitats that beaches represent for these animals are highlighted. I start by explaining the ecological importance of Crustaceans in sandy beach ecosystems worldwide. The most abundant groups in temperate beaches and their main environmental stressors are afterwards presented. Finally, specific management and conservation strategies, employing Talitrids and Tylids as ecological indicators of environmental quality condition in beaches from temperate latitudes, are identified.

1. INTRODUCTION

Sandy beaches are the dominant ecosystem in worldwide coastlines and represent a valuable natural resource since they offer several goods and services, some of which with economical relevance (e.g., fishing areas for fish and shellfish, bait harvesting, coastal protection, recreation and relaxation activities), as well as unique feeding and/or reproduction grounds for some iconic species. Also, they house an exceptional biota where the meiofauna, living in the interstices between the sand grains, is the most conspicuous and diverse component. However, only a few species present abundant populations in sandy beaches, especially among the macrofauna of the supralittoral zone and, compared to other marine and littoral systems, diversity is relatively low, causing a high sensitivity to changes.

In conditions of high exposure to the winds and wave action, animals with a high degree of mobility and a strong and quick burrowing ability, like the Crustaceans, are better adapted to the demanding environment (McLachlan and Brown, 2006) that shapes the macrofaunal community of a sandy beach. In this scenario other less fit taxa, like molluscs or polychaetes, will be less abundant or even excluded. In most cases, beaches are unique habitats for a large number of these animals, which are not replaceable once their environmental quality is lost or even seriously damaged.

Today, sandy beaches are one of the most endangered coastal environments. As a result of the economic progress, coastal development has been intensified in the last decades, resulting in widespread modifications in these ecosystems (Defeo et al. 2009). Besides this, since the 1950s, an enormous growth in beach tourism has occurred (Caffyn et al. 2002), bringing more and more tourists to these coastal systems every year. This growth has only been possible due to the introduction of several touristic and urbanistic facilities in the beach and its surrounding areas, with the inherent costs that this represents for the ecological balance of these natural areas (see for instance Davenport and Davenport, 2006; Bessa et al. 2013; Gonçalves et al. 2013). Unfortunately, mass tourism in coastal areas is nowadays an undeniable reality in several parts of the globe.

On the other hand, sandy beaches are also subjected to other environmental problems, mostly at a global scale and related with human activities as well. These include beach erosion and beach narrowing, associated with an increase in storms occurrence and with the rise in sea level due to global warming (Doney et al. 2012), and changes in the groundwater table within the beach and dune systems by depressing or elevating groundwater levels, which conduct to significant impacts on the beach face (erosion, deposition of sand) (McLachlan and Brown, 2006). Under all these circumstances, it is extremely important to understand the responses of these ecosystems to their distinct pressures and stressors, if ecological condition assessment, management and conservation strategies are to be developed and successfully implemented.

In this commentary the relevance of using specific groups of Crustaceans as ecological tools in management and conservation strategies for temperate sandy beaches is discussed. Also, the benefits of these strategies for the conservation of the unique habitats that beaches represent for these animals are highlighted. I start by explaining the ecological importance of Crustaceans in sandy beach ecosystems worldwide. The most abundant groups in temperate beaches and their main environmental stressors are afterwards presented. Finally, specific management and conservation strategies, employing Talitrids and Tylids as ecological indicators of environmental quality condition in beaches from temperate latitudes, are identified.

2. The Importance of Crustaceans in Sandy Beach Ecosystems

Macrofauna is a fundamental element in the structure and function of the beach ecosystem. Since primary macroproducers, such as macrophytes and macroalgae, are usually absent due to substrate instability and wave action in intertidal areas, macrofauna relies mostly on allochthonous food inputs, mainly in the form of stranded macrophytes and detritus (particulate organic matter) on a daily basis (see Colombini and Chelazzi, 2003 for a *review* on the subject). These are mainly brought ashore by tides, although they may also be blown by the winds from a terrestrial origin. A large proportion of the macrofauna of exposed sandy beaches, namely supralittoral amphipods, isopods and insects, is therefore composed of scavengers, dependent of the wrack subsidies for food, although other functional components of the macrofauna might also be dependent of the wrack for the provision of shelter or adequate habitats (Gonçalves and Marques, 2011). Consequently, macrofauna contributes for wrack consumption and degradation and, simultaneously, serves as prey for several top predators such as fishes or shorebirds, occupying a central position in the beach food webs.

As previously mentioned, Crustaceans are the dominant faunal group in the macrofaunal communities of exposed sandy beaches, although molluscs, polychaetes and, particularly insects, are also common (McLachlan and Brown, 2006). Also, they are the most widespread and common invertebrates in all tidal levels of worldwide sandy beaches (McLachlan and Jaramillo, 1995), namely on the supralittoral and on the intertidal areas of the beach. Peculiar features like complex zonation patterns related with the tidal gradient, fluctuations due to tidal and other types of migrations (Knox, 2001), and behavioural plasticity to these rapidly changing environments are common on beach macrofaunal inhabitants (Brown, 1996), especially on Crustaceans. Common Crustaceans in exposed sandy beaches include amphipods, such as Talitrids and Haustoriids, isopods, especially Tylids and Cirolanids, Ocypodid crabs and Callianassid shrimps.

3. Prominent Crustaceans in Temperate Sandy Beaches

In temperate latitudes, amphipods from the family Talitridae and isopods from the families Tylidae and Cirolanidae, stand out as prominent inhabitants of sandy beaches (Figure 1), due to their abundance. Talitrids and Tylids are mainly dominant in the supralittoral zone of the beach, although some species of Talitrids also occupy the intertidal mostly on a seasonal basis. Cirolanids, on the other hand, are characteristic of the intertidal zone. However, sand hoppers (Talitrids) are by far the most conspicuous animal to the naked eye in temperate beaches, and possibly, often the only animals that common beach users recognize and associate with the beach environment.

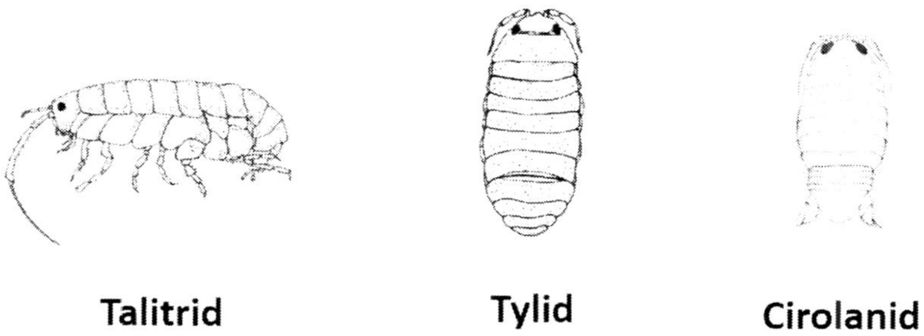

Figure 1. Schematic representation of the most abundant Crustacean families in the supralittoral and the intertidal zones of exposed sandy beaches from temperate regions.

Regarding Talitrid amphipods, common genera include *Talitrus*, *Orchestia*, and *Talorchestia*, which are scavengers or herbivorous, and often live associated with the wrack deposits found on the upper realms of the beach (McLachlan and Brown, 2006). Talitrids have complex orientation and migration skills, moving actively on diel, tidal and also seasonal scales to avoid inappropriate environmental conditions, such as submersion, and to locate stranded wrack deposits (see for instance Scapini and Dugan, 2008; Gonçalves et al. 2009). The species *Talitrus saltator* has been widely studied, especially on Atlantic and Mediterranean shores, and proposed as an ecological indicator species, namely for disturbances derived from human activities (Ugolini et al. 2008; Gonçalves et al. 2013; Bessa et al. 2013). Other species, such as *Talorchestia brito*, *Talorchestia quoyana* or *Orchestia gammarellus* are also common inhabitants in temperate sandy beaches (see for instance Gonçalves et al. 2003; Colombini et al. 2013).

The genus *Tylos* comprises the vast majority of the species in the family Tylidae, and presents a circumglobal distribution since it is not restricted to temperate regions. Like Talitrids, these animals are mainly scavengers or herbivorous, and are wrack-associated fauna, dominant on the supralittoral zone where they often standout as abundant populations in macrofaunal assemblages. In temperate sandy beaches, *Tylos europaeus* forms abundant populations, extensively recorded in Atlantic and Mediterranean coastlines, especially of Europe (Deidun et al. 2003; Gonçalves et al. 2005; Gonçalves et al. 2009). Other species, such as *Tylos sardous*, and especially, *Tylos ponticus* are also often present in temperate sandy beaches (*e.g.,* Montesanto et al. 2014).

In the family Cirolanidae, the widespread *Eurydice* and the apparently ubiquitous *Excirolana* are prominent genus mainly in the intertidal zones of several worldwide coasts. Common species include for instance *Eurydice pulchra* and *Eurydice affinis* in Europe, *Excirolana braziliensis*, with a wide Pan-American distribution from tropical to temperate latitudes (*e.g.,* Cardoso and Defeo, 2003), and *Excirolana hirsuticauda* in temperate beaches of the vast Chilean coast (*e.g.,* Contreras and Jaramillo, 2003). These Cirolanids are mainly scavengers or predators, dependent of wrack deposits or on the presence of other faunal components that serve as their preys.

4. Talitrids and Tylids as Ecological Indicators in Temperate Latitudes

As mentioned in the Introduction section, sandy beaches are today one of the most endangered coastal ecosystems. The supralittoral and the intertidal zones of the beach are clearly the areas needing priority intervention regarding environmental assessment and management programs. Besides the environmental problems that threaten the worlds' coastlines at a global scale (e.g., climatic changes, rise of the sea level, beach erosion and narrowing), several anthropogenic disturbances may also jeopardize beaches ecological states.

Increases in touristic and recreational activities on the beach produce several kinds of impacts which may cause macrofaunal populations to decline. For instance, intense human trampling and also mechanical trampling by off-road vehicles (used in recreational activities), can damage the animals, destroy their burrows or even fatally crush them, as demonstrated for instance by Węslawski et al. (2000) and by Veloso et al. (2006). Beach grooming with heavy machinery, on a regular basis or even only seasonally, can pose several problems to macrofaunal animals, especially to wrack-associated taxa. Besides removing important food resources, the physical impact that heavy machines introduce in the sand environment, where these animals are burrowed during the day, causes for instance a decrease or even the disappearance of this macrofauna, but also of some of their predators such as wading birds (Llewellyn and Shackley, 1996; McLachlan and Brown, 2006).

The construction of touristic, recreational and some urban associated facilities is often responsible for coastal dune degradation and destruction, and common for instance on highly developed and touristic European coasts. This kind of impact disturbs coastal dune biota, and jeopardizes some of the important ecological services that the beach/dune system provides. Furthermore, coastal dune degradation and destruction may also disturb sandy beach macrofauna. Supralittoral organisms, like Talitrids, are particularly vulnerable due to their ability to make seasonal movements on the beach/dune system, using the dunes for overwintering or even to avoid extreme high temperatures in summer (e.g., Marques et al. 2003). This kind of activity was also proposed to interfere with the movements of other sandy beach crustaceans between the beach and the dune, such as in the widespread burrowing ghost crabs at the Australian beaches (Barros, 2001).

Some macrofaunal species form abundant populations on exposed beaches, often functioning as the driving forces of the community and shaping both the community's structure and function. Undeniably, these abundant species are key elements since they have a relevant role on the equilibrium of the ecosystem (see for instance Gonçalves et al. 2009; Gonçalves and Marques, 2011). Removal or depletion of a key species, as well as strong abnormal population growths, have profound effects in the structure of the community and on the ecosystems equilibrium. Because this equilibrium is dynamic, changes in the ecosystem, caused for instance by human actions, will also be reflected in key species populations.

Regarding management, key species are important indicators of sustainability (Scapini and Morgan, 2002), but may also function as indicators of environmental changes in the dynamic and complex conditions that characterize the exposed sandy beach ecosystem (Gonçalves et al. 2013). From an ecological management point of view, it is also extremely important that the key species is simultaneously widespread and nevertheless fairly sensitive to the human pressures on the beach.

In temperate sandy beaches, Talitrid amphipods and Tylid isopods can be used as *ecological indicators* of environmental quality condition, since they frequently present key species in the macrofaunal communities of exposed beaches. This type of indicators are fundamental tools in environmental assessment, monitoring, management and conservation programs, crucial for the integrity and sustainability of the beach biota and their habitats. In fact, the

Talitrids *Talitrus saltator* and *Talorchestia brito*, as well as the Tylid *Tylos europaeus*, were recognized as key species in the western coast of Portugal by Gonçalves et al. (2009) and are believed to have similar roles in other exposed sandy shores of Europe and even North Africa, where they often form abundant populations. *Talitrus saltator*, in particular, has already been considered an important ecological indicator of disturbances derived from human activities by several authors (*e.g.,* Ugolini et al. 2008; Gonçalves et al. 2013; Bessa et al. 2013).

Studying the bioecology of Talitrid and Tylid populations in undisturbed or low impacted sandy beaches produces important sets of ecological data that function as reference attributes in future assessments, allowing to evaluate the impact of specific disturbances occurring in the ecosystem. Bioecological features such as population dynamics (densities, mortality, reproduction and recruitment), reproductive biology (extension of the reproductive period, fecundity and voltinism) and secondary production estimates (growth and elimination productions, average annual biomass and population turnovers) are particularly relevant and useful. For instance, oscillations in the reproductive strategies or in the population dynamics of a given species may reflect the occurrence of a specific environmental disturbance. This type of approach has been supported for instance by Marques and Anastácio (2002) and by Gonçalves et al. (2013). Also, comparative studies of distinct populations on local and macro scale scenarios allow the detection of variations on population bioecology resulting from the adaptability of the populations to the local scale environments or even to large-scale latitudinal variations, as proposed by Gonçalves et al. (2013).

Population dynamics models, capable of simulating the dynamics of a given Talitrid or Tylid key species population, is another powerful tool. It allows the ecological manager to simulate multiple scenarios that might reflect potential changes in environmental conditions, namely disturbances induced by human activities. For instance, scenarios such as significant increases in human trampling – caused by high numbers of beach users – or significant decreases in food and refuge availability – caused by intense grooming activities on the beach – can be rapidly simulated. The outputs of these simulations will reveal the most probable responses of the population, allowing to anticipate community changes and allowing the implementation of preventive management or mitigation actions. However, this tool is clearly underdeveloped, most probably because large expertise and profound baseline knowledge regarding the dynamics of natural populations are needed. Also, the several steps necessary to construct, calibrate and validate the model regarding prediction and replication is a highly time-consuming process.

Since the macrofaunal communities of sandy beaches can respond relatively rapid to natural and anthropogenic disturbances (Jones et al. 2008; Fanini et al. 2009; Schlacher et al. 2012), community structural attributes that are easily visualised, like for instance the abundance of common species, can be used to monitor short and long term responses. However, previous information upon what to expect of the structural attributes on undisturbed (or as less disturbed possible) communities are needed, if potential drivers of community changes are to be identified. This approach has been often used by sandy beach ecologists to assess for example the effects of some specific human activities on macrofaunal communities (see for instance Fanini et al. 2009; Leewis et al. 2012; Bessa et al. 2013).

The structure and the function of ecosystems are deeply linked and cannot be dissociated from each other. Therefore, community functional attributes, like for instance trophic structure (functional groups), secondary production and standing stock estimates of the most abundant species, such as Talitrids and Tylids, may also represent interesting tools to use on beach management and assessment practices. Nevertheless, functional traits still remain largely unexplored by beach ecologists.

Conclusion

Abundant macrofaunal populations of Crustaceans, that stand out as key elements in the structure and function of their communities, hold large potential as ecological indicators of the environmental quality condition and of the environmental disturbances occurring in exposed sandy beach ecosystems. In the present scenario of increasing detrimental impacts on the beach environment, but also of the degradation and regression of the coastlines in several parts of the world, the responses of these indicators to environmental disturbances derived from human activities are especially relevant.

In temperate beaches, and partially due to the amount of ecological knowledge already produced about these Crustaceans, Talitrid amphipods and Tylid isopods are the most promising groups useful as ecological indicators. However, Cirolanid isopods may also have similar roles, but a larger investment in studying and understanding their bioecology and their responses to human derived disturbances is mandatory, especially for the genus *Eurydice*.

Bioecological reference attributes, about what to expect of population ecology in undisturbed or low impacted ecosystems, are important to understand local and also macroscale variations in the bioecology of a key species. This functions as baseline knowledge against which it is possible to detect distinct bioecological responses to other sources, such as human activities, and to stablish these links by using appropriate statistical techniques. Community structural attributes are another important assessment and monitoring strategy, most often used in several studies targeted to understand the impacts derived from human activities on the beach biota. The use of this strategy focused on Talitrid or Tylid species may allow to detect the responses of the ecosystem to some specific stressors more rapidly.

Population biological and/or ecological features and community structural attributes have been the prevalent tools on sandy beach assessment, monitoring and management practices, although monitoring and management strategies for these ecosystems are still underdeveloped. The construction of population dynamics models for Talitrid or Tylid key species, however, would be an asset for environmental management and beach conservation, since these are very powerful simulation tools. The beach manager/ecologist can simulate a multitude of distinct environmental scenarios and forecast the population's responses to the stressor introduced in the simulation. Also, beach grooming is a common practice in beaches heavily explored by tourism, and it has significant negative impacts on the macrofauna and on the community function. Altogether, these aspects reinforce the need for the development of both population dynamics models and of ecological indicators for sandy beach assessment and management based on community functional attributes, promising subjects for further investigation.

REFERENCES

Barros, F. (2001). Ghost crabs as a tool for rapid assessment of human impacts on exposed sandy beaches. *Biological Conservation*, *97*, 399-404.

Bessa, F., Cunha, D., Gonçalves, S. C. & Marques, J. C. (2013). Sandy beach macrofaunal assemblages as indicators of anthropogenic impacts on coastal dunes. *Ecological Indicators*, *30*, 196–204.

Brown, A. C. (1996). Behavioural plasticity as a key factor in the survival and evolution of the macrofauna on exposed sandy beaches. *Revista Chilena de História Natural*, *69*, 469-474.

Caffyn, A., Prosser, B. & Jobbins, G. (2002). A Framework for the analysis of socio-economic impacts on beach environments. In: F. Scapini (Ed.) *Baseline research for the integrated sustainable management of Mediterranean sensitive coastal ecosystems. A manual for coastal managers, scientists and all those studying coastal processes and management in the Mediterranean*, (37-50). Florence, Istituto Agronomico per l'Oltremare.

Cardoso, R. S. & Defeo, O. (2003). Geographical patterns in reproductive biology of the Pan-American sandy beach isopod *Excirolana braziliensis*. *Marine Biology*, *143*, 573-581.

Colombini, I. & Chelazzi, L. (2003). Influence of marine allochthonous input on sandy beach communities. *Oceanography and Marine Biology: an Annual Review*, *41*, 115-159.

Colombini, I., Fallaci, M., Gagnarli, E., Rossano, C., Scapini, F. & Chelazzi, L. (2013). The behavioural ecology of two sympatric talitrid species, *Talitrus saltator* (Montagu) and *Orchestia gammarellus* (Pallas) on a Tyrrhenian sandy beach dune system. *Estuarine, Coastal and Shelf Science*, *117*, 37-47.

Contreras, H. & Jaramillo, E. (2003). Geographical variation in natural history of the sandy beach isopod *Excirolana hirsuticauda* Menzies (Cirolanidae) on the Chilean coast. *Estuarine, Coastal and Shelf Science*, *58S*, 117-126.

Davenport, J. & Davenport, J. L. (2006). The impact of tourism and personal leisure transport on coastal environments: a review. *Estuarine, Coastal and Shelf Science*, *67*, 280-292.

Defeo, O., McLachlan, A., Schoeman, D. S., Schlacher, T. A., Dugan, J., Jones, A., Lastra, M. & Scapini, F. (2009). Threats to sandy beach ecosystems: a review. *Estuarine, Coastal and Shelf Science*, *81*, 1–12.

Deidun, A., Azzopardi, M., Saliba, S. & Schembri, P. J. (2003). Low faunal diversity on Maltese sandy beaches: fact or artefact? *Estuarine, Coastal and Shelf Science*, *58*, 83-92.

Doney, S. C., Ruckelshaus, M., Duffy, J. E., Barry, J. P., Chan, F., English, C. A., Galindo, H. M., Grebmeier, J. M., Hollowed, A. B., Knowlton, N., Polovina, J., Rabalais, N. N., Sydeman, W. J. & Talley, L. D. (2012). Climate change impacts on marine ecosystems. *Annual Review of Marine Science*, *4*, 11-37.

Fanini, L., Marchetti, G. M., Scapini, F. & Defeo, O. (2009). Effects of beach nourishment and groynes building on population and community descriptors of mobile arthropodofauna. *Ecological Indicators*, *9*, 167–178.

Gonçalves, S. C., Marques, J. C., Pardal, M. A., Bouslama, M. F., El Gtari, M. & Charfi-Cheikhrouha, F. (2003). Comparison of *Talorchestia brito* (Amphipoda, Talitridae) biology, dynamics, and secondary production in Atlantic (Portugal) and Mediterranean (Tunisia) populations. *Estuarine, Coastal and Shelf Science*, *58*, 901-916.

Gonçalves S. C., Pardal, M. A., Cardoso, P. G., Ferreira, S. M. & Marques, J. C. (2005). Biology, population dynamics and secondary production of *Tylos europaeus* (Isopoda, Tylidae) on the western coast of Portugal. *Marine Biology*, *147*, 631-641.

Gonçalves, S. C., Anastácio, P. M., Pardal, M. A., Cardoso, P. G., Ferreira, S. M. & Marques, J. C. (2009). Sandy beach macrofaunal communities on the western coast of Portugal – Is there a steady structure under similar exposed conditions? *Estuarine, Coastal and Shelf Science*, *81*, 555-568.

Gonçalves, S. C. & Marques, J. C. (2011). The effects of season and wrack subsidy on the community functioning of exposed sandy beaches. *Estuarine, Coastal and Shelf Science*, *95*, 165-177.

Gonçalves, S. C., Anastácio, P. M. & Marques, J. C. (2013). Talitrid and Tylid crustaceans bioecology as a tool to monitor and assess sandy beaches' ecological quality condition. *Ecological Indicators*, *29*, 549-557.

Jones, A. R., Murray, A., Lasiak, T. A. & Marsh, R. E. (2008). The effects of beach nourishment on the sandy-beach amphipod *Exoediceros fossor*: impact and recovery in Botany Bay, New South Wales, Australia. *Marine Ecology*, *29*, 28–36.

Knox, G. A. (2001). The ecology of seashores. Ed: M. J. Kenish. CRC Press LLC, Boca Raton.

Leewis, L., van Bodegom, P. M., Rozema, L. & Janssen, G. M. (2012). Does beach nourishment have long-term effects on intertidal macroinvertebrate species abundance? *Estuarine, Coastal and Shelf Science*, *113*, 172–181.

Llewellyn, P. J. & Shackley, S. E. (1996). The effects of mechanical beach cleaning on invertebrate populations. *British Wildlife*, *7*(3), 147-155.

Marques, J. C. & Anastácio, P. M. (2002). Integration and interpretation of ecological data at population level. In: F. Scapini (Ed.) *Baseline research for the integrated sustainable management of Mediterranean sensitive coastal ecosystems. A manual for coastal managers, scientists and all those studying coastal processes and management in the Mediterranean* (155-164). Florence, Istituto Agronomico per l'Oltremare.

Marques, J. C., Gonçalves, S. C., Pardal, M. A., Chelazzi, L., Colombini, I., Fallaci, M., Bouslama, M. F., El Gtari, M., Charfi-Cheikhrouha, F. & Scapini, F. (2003). Comparison of *Talitrus saltator* (Amphipoda,

Talitridae) biology, dynamics, and secondary production in Atlantic (Portugal) and Mediterranean (Italy and Tunisia) populations. *Estuarine, Coastal and Shelf Science*, *58S*, 127-148.

McLachlan, A. & Brown, A. C. (2006). The ecology of sandy shores (second edition). California, Elsevier.

McLachlan, A. & Jaramillo, E. (1995). Zonation on sandy beaches. *Oceanography and Marine Biology. An Annual Review*, *33*, 305-335.

Montesanto, G., Deidun, A., Sciberras, A., Sciberras, J. & Bianca Maria Lombardo, B. M. (2014). Current distribution of two species of *Tylos* (Isopoda: Oniscidea) in the Central Mediterranean and the influence of beach sand grain-size parameters. *Journal of Crustacean Biology*, *34*(1), 47-53.

Scapini, F. & Morgan, E. (2002). Bioassays for estimation of beach stability and ecosystem quality. In: F. Scapini (Ed.), *Baseline research for the integrated sustainable management of Mediterranean sensitive coastal ecosystems. A manual for coastal managers, scientists and all those studying coastal processes and management in the Mediterranean*, (120-122). Florence, Istituto Agronomico per l'Oltremare.

Scapini, F. & Dugan, J. (2008). Sun and landscape orientation in adult and juvenile sandhoppers *Orchestoidea tuberculata* (Amphipoda, Talitridae) from two beaches in south central Chile. *Marine Ecology*, *29*, 19–27.

Schlacher, T. A., Noriega, R., Jones, A. & Dye, T. (2012). The effects of beach nourishment on benthic invertebrates in eastern Australia: Impacts and variable recovery. *Science of the Total Environment*, *435-436*, 411–417.

Ugolini, A., Ungherese, G., Somigli, S., Galanti, G., Baroni, D., Borghini, F., Cipriani, N., Nebbiai, M., Passaponti, M. & Focardi, S. (2008). The amphipod *Talitrus saltator* as a bioindicator of human trampling on sandy beaches. *Marine Environmental Research*, *65*, Issue 4, 349-357.

Veloso, V. G., Silva, E. S., Caetano, C. H. S. & Cardoso, R. S. (2006). Comparison between the macroinfauna of urbanized and protected beaches in Rio de Janeiro state (Brazil). *Biological Conservation*, *127*, 510-515.

Węslawski, J. M., Stanek, A., Siewert, A. & Beer, N. (2000). The sandhopper (*Talitrus saltator*, Montagu 1808) on the Polish Baltic coast. Is it a victim of increased tourism? *Oceanological Studies*, vol *XXIX*, no. 1, 77-87.

Index

A

B

C

D

E

F

G

H

N

O

P

R

S

T

V

W

X

Y

Z

V

W

X

Y

Z